# Google Earth Forensics

## Using Google Earth Geo-Location in Digital Forensic Investigations

# Google Earth Forensics

## Using Google Earth Geo-Location in Digital Forensic Investigations


Michael Harrington

Michael Cross

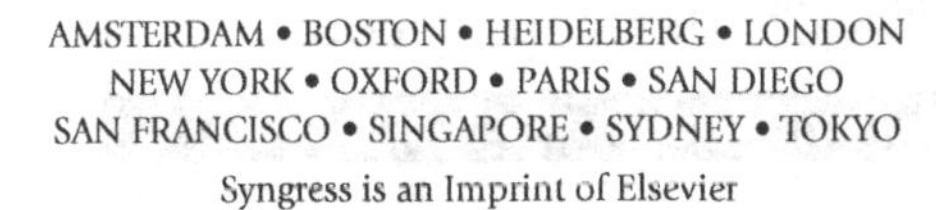


SYNGRESS

Acquiring Editor: Chris Katsaropoulos
Editorial Project Manager: Benjamin Rearick
Project Manager: Surya Narayanan Jayachandran
Designer: Matthew Limbert

Syngress is an imprint of Elsevier
225 Wyman Street, Waltham, MA 02451, USA



**British Library Cataloguing-in-Publication Data**
A catalogue record for this book is available from the British Library

**Library of Congress Cataloging-in-Publication Data**
A catalog record for this book is available from the Library of Congress

ISBN: 978-0-12-800216-2

For information on all Syngress publications
visit our website at http://store.elsevier.com/

# Contents

**BIOGRAPHY** ........ vii

**CHAPTER 1** Google Earth Basics ........ 1
What is google earth? ........ 1
Google earth for forensics ........ 2
Flavors of google earth ........ 4
Installing google earth on your computer ........ 8

**CHAPTER 2** Using Google Earth ........ 11
The google earth UI ........ 11
Navigation ........ 19
Views ........ 22
Tours ........ 26
Configuration ........ 28

**CHAPTER 3** GPS, GIS, and Google Earth ........ 35
Understanding GPS ........ 35
Understanding GIS ........ 42
Geo-location information in pictures ........ 45

**CHAPTER 4** KML/XML/HTML ........ 49
Markup Languages and Google Earth ........ 49
Using HTML in google earth ........ 51
What is KML? ........ 53
XML ........ 56
KML revisited ........ 59
Learning more about markup languages ........ 67

**CHAPTER 5** Digital Forensics 101 ........ 69
What is digital forensics? ........ 69
Tools for recovering evidence ........ 74
Do you really want to do this? ........ 77
Organizing your case ........ 77
Understanding what you are looking at ........ 80

**CHAPTER 6** Working a Case .......... 83
The practical application of google earth forensics .......... 83
Acquiring from a GPS unit .......... 85
Annotating a crime scene .......... 88
Views and camera angles .......... 101
Legends, logos, and banners .......... 103
Creating a tour of the crime scene .......... 107
Distributing your work in google earth .......... 109

**INDEX** .......... **111**

# Biography

Michael Cross (MCSE, MCP+I, CNA, Network+) is an Internet specialist/computer forensic analyst with the Niagara Regional Police Service (NRPS). He performs computer forensic examinations on computers involved in criminal investigation. He also has consulted and assisted in cases dealing with computer-related/Internet crimes. In addition to designing and maintaining the NRPS Web site at www.nrps.com and the NRPS intranet, he has provided support in the areas of programming, hardware, and network administration. As part of an information technology team that provides support to a user base of more than 800 civilian and uniform users, he has a theory that when the users carry guns, you tend to be more motivated in solving their problems.

Michael also owns KnightWare (www.knightware.ca) that provides computer-related services such as Web page design, and Bookworms (www.bookworms.ca), where you can purchase collectibles and other interesting items online. He has been a freelance writer for several years, and he has been published more than three dozen times in numerous books and anthologies. He currently resides in St. Catharines, Ontario, Canada, with his lovely wife, Jennifer, his darling daughter, Sara, and charming son, Jason.

Michael Harrington is a former law enforcement officer with over 10 years of experience in digital forensics. He lectures on mobile forensics around the world and has been involved in various forensic projects including Pandora's Box and WOLF. Michael has been published in the Thomas J Cooley Law Journal and on Forensic Focus. He also writes on the subject of mobile forensics at http://mobileforensics.wordpress.com/.

CHAPTER 1

# Google Earth Basics

**INFORMATION IN THIS CHAPTER:**

- What is Google Earth?
- Google Earth for Forensics
- Flavors of Google Earth
- Installing Google Earth

## WHAT IS GOOGLE EARTH?

Google Earth (GE) is a tool that provides you with the ability to view the planet through a virtual globe, and tunnel down to examine more detailed information. Using it, you can navigate through satellite images, aerial photography, and even views of street level imagery and 3D models of the world. This not only includes landmasses on Earth, but other locations like oceans, the moon, Mars, and outer space. Features in Google Earth even allow you to take tours of locations, or fly across locations using a flight simulator.

Looking at some of these features, you might think that Google Earth is just a novelty or some kind of toy, but that is not the case. By entering in an address or coordinates of a location, GE will display a map that includes the labeled position of the place you are searching for. You can then zoom in to see 3D structures or actual photos of a location. As we will see in the chapters that follow, you can also view areas of the earth using custom maps or overlays, which contain data imported from GPS units, other devices, or files you have created. The real world, practical applications for this tool are varied and sometimes amazing.

### How Google Earth is Being Used

As a resource, it is often only limited by your resourcefulness. For years, Google Earth has been used by teachers creating lesson plans involving geography



Google Earth Forensics

and location. Through this application, students can see population densities, learn about migration, and how cultures have evolved and interacted in different locations. Other examples of the usefulness of this tool include:

- In 2014, aboriginal groups began using Google Earth to map First Nations territories in Canada. Some of its uses will be to compare relationships, track environmental changes, and resolve potential land-claim issues [1].
- The U.S. Fish and Wildlife Service provides data on wetlands that can be displayed in Google Earth [2].

Police throughout the world have used Google Earth in a variety of ways, from investigating crimes to sharing information with the public. Crime analysts in law enforcement agencies gather data from police reports and/or other sources, and may have this information made available through Google Earth. For example, the Shawnee Police Department in Kansas provides data that can be loaded into GE to see locations where robberies, auto thefts, vandalism and other crimes have taken place [3]. In another example, Sheriff's deputies in Humboldt County, California pulled over a man with approximately $63,000 in marijuana, and used the coordinates in his GPS device to find several fields of pot. The local coordinates were entered into Google Earth, allowing them to display the locations quickly and gain an understanding of the terrain.

As you might expect, if Google Earth is used by police, the criminals are also probably using it. In 2009, Tom Berge used it to review aerial photographs to find buildings that had lead roof tiles. Over a 6-month period, he would find targets with Google Earth by searching for darker than normal roofs. He would then go to a location where he would climb onto the roof, and steal the tiles so they could be sold for scrap metal [4]. Another example of GE being used for criminal activity occurred in 2014, when burglars were found to be using it to scout the best way in and out of houses they were breaking into [5].

## GOOGLE EARTH FOR FORENSICS

Forensics is the use of scientific or technological techniques to investigate and establish facts. In a criminal case, the facts you are looking for will be evidence of how a crime was committed and who was responsible. In looking at traditional methods, an investigator would visit a crime scene and gather fingerprints, fibers, and take photographs. Throughout a process of preserving the crime scene and identifying, gathering and examining evidence, information is carefully documented. This is used in the hopes of understanding what occurred, and so that it may be used to identify, arrest and convict the person(s) responsible. Even though it is physical evidence that is being handled, the same basic principles apply to digital forensics.

Digital forensics is a newer branch of forensics, in which evidence is gathered from computers and other devices capable of storing digital information. This not only includes data stored on a computer's hard disk, but geo-location information, pictures and other data stored in mobile phones, tablets, GPS units and other devices. As we will discuss in Chapter 5, just like fingerprints and other physical evidence is investigated by following best practices and procedures, digital forensics also follows a process of seizing, acquiring, analyzing, and reporting what evidence is found.

Digital forensics differs from traditional forensics in how it is used and who is using it. A criminal investigation will generally be conducted by members of law enforcement, but digital forensics may also be used by organizations and individuals. A member of the IT department may be the first responder to a breach in security (such as a hacking attempt), discover illegal photos on a computer, or involved in internal investigations of other employees. In other circumstances, such as private investigation, the data on a device might be examined to determine if a person had conducted themselves in an inappropriate or illegal manner. Regardless of the situation, an investigator may use the same software and hardware as police, and should follow the same best practices and procedures that law enforcement use. Not only do these methods prevent evidence from being compromised, but they will prevent the evidence from being challenged and dismissed if the case needs to go to civil or criminal court.

While we will discuss a number of tools used to acquire data from a device, the location-specific information gathered often is not meaningful until it is analyzed. The raw data is useful, but if it is a series of coordinates referencing a location on the Earth, what does that tell you? After all, if I tell you I found the coordinates 38.8977°N, 77.0366°W on a cell phone, it is not as significant as showing you that this is the location of the White House.

Using Google Earth, you can search and display location-specific information in a way that is more telling than the raw data. Using it as a forensic tool, you can so such things as:

- Import data from mobile devices and GPS units to determine a route that was taken, or locations that a person visited
- Determine the location where a photo was taken using geo-location information stored in a digital picture
- Create maps that display locations that a person visited, and movies that covey location-based information in a compelling format for investigations and court

Throughout this book, we will discuss these and other ways Google Earth can be used and the methods involved in acquiring, analyzing and reporting

**Table 1.1** Google Earth for Web Requirements

| Operating System | Browser |
|---|---|
| Microsoft Windows Vista or higher | Google Chrome 5.0-37.0 (32-bit)<br>Internet Explorer 7-9, 10-11 (in compatibility mode (32-bit))<br>Firefox 2.0 or higher<br>Flock 1.0 or higher |
| Apple Mac OS X 10.6 or higher | Google Chrome 5.0-37.0<br>Safari 3.1 or higher<br>Firefox 3.0 or higher |

geo-location information. However, before you get a taste of this, let us look at the flavors of Google Earth available to you.

## FLAVORS OF GOOGLE EARTH

When you visit the Google Earth site (www.google.com/earth), you will find there are several variations of the product for different platforms and different uses. These are:

- Web
- Mobile
- Desktop
- Enterprise

As we will see in the sections that follow, each of these versions allow you to view and navigate through 3D maps, satellite images, and view Earth data. Not all of them will be equally useful to your needs, so it is important to consider how you plan to use the tool, and what you will be using to run it.

### Google Earth for Web

Originally released in 2008, the Web version is a free plug-in that installs on your browser, and allows you to use features of Google Earth through Web pages. Web developers can place a version of GE on Web pages, displaying custom maps or integrating features into online applications that communicate with a free JavaScript API (Application Programming Interface) that Google provides. When you visit such a site, the page loads and the plug-in can display a 3D globe, custom maps, and other Google Earth content.

The plug-in is available on the Google Earth site. By visiting the plug-in page at http://earth.google.com/plugin.html, you can download and install the plug-in on any of the supported browsers and operating systems (Table 1.1).

It is important to note that if you are trying to install the plug-in on Firefox, it will not install while Firefox is running. As such, you need to download the plug-in, shutdown Firefox, and then run the installation.

### Google Earth for Mobile Devices

The mobile app for Google Earth was also released in 2008, and allows you to access Google Earth features through a smartphone, tablet or other mobile devices. Versions of the app are available for iOS and Android devices, and available to download through app stores like Google Play and iTunes. In 2012, Google also provided the ability to view custom content on the mobile version of GE, allowing you to view any custom maps or overlays available on a page by clicking the link to an associated KML file (which we will discuss in detail in Chapter 3). Once you click on the link to a Google Earth file, the mobile app will open and automatically launch and load the custom map.

The app has features that are remarkably useful to mobile users, such as the ability to view public transit and traffic information. The traffic view overlays colored lines on a map to tell you the estimated speed that traffic is flowing, and icons indicating road closures and other conditions. To get an idea of this feature on a computer, you can visit Google Maps (http://maps.google.com) and type "Traffic Near" followed by the name of a location into the search box. Similarly, the public transit view overlays transit lines on a map to tell you about public transportation in a particular area. You can also see this in Google Maps by searching for a location, and when the map the area appears, typing "Transit" into the search box.

### Google Earth Enterprise

In addition to these two versions, there is also an Enterprise edition, which is designed for larger organizations. Using Google Earth Enterprise (GEE), a company can create and store imagery on their own server infrastructure, and make it available to users. The maps can be published and viewed with any of the previously mentioned versions of Google Earth, or through a custom application that uses the Google Maps API.

GEE is actually a package of software that resides on client and server machines, and is made up of several components:

- Fusion, which is used to combine images, terrain, KML and other data into a globe or map
- Server Software, which is used to host the globe and maps provided to users
- Client, which allows users to view the globe and maps
- Google Earth API, which allows developers to incorporate Google Earth features into Web applications and pages, allowing your company's data is viewed in a way that's customized to your organization's needs

GEE is used to distribute geographic information across a wide audience of users in an organization. For example, an advertising company might use it to create a custom map of current customers, which the sales people could use to view information on clients. A police department might use it to record the location of different types of crimes, which could then be used as a reference for high-crime areas and to see where clusters of certain crimes are committed. Such applications of the tool could be helpful in determining where additional people need to be deployed to properly service an area.

The kneejerk reaction might be to buy the Enterprise version, since those types of versions generally have the most features. For forensics purposes, GEE is probably more than you need. Google Earth is used in such circumstances to examine and report geographic data, and generally not to publish evidence to a large audience. As such, in discussing Google Earth in this book, we will be discussing the Free or Pro versions of the product.

## Free vs. Pro Desktop Versions

There are two versions of Google Earth that you can install on your computer.

- Google Earth, which is a free version of the application
- Google Earth Pro, which (at the time of this writing costs $399 per year), and provides additional tools for business users. If you are unsure if you want to purchase the Pro version, a 7-day trial version is available.

The free version of Google Earth is intended for home or personal use, and is often the best choice for others to view the findings of your investigation. Without having to purchase a copy of the program, a person can install the free version on their desktop and view satellite and aerial imagery, as well as custom map data. You can also import and manually geo-locate GIS images, and save or print the information you're viewing as screen resolution images. Unlike the Pro version, the free copy of this product contains in-product ads. This should not be a big surprise. After all, if you are not buying it, Google has to make their money somehow.

Google Earth Pro is designed for commercial use, and provides the same features as the free version with additional features added. Using it, you can measure areas of a polygon or circle on a map, and print or save the information as high-resolution images. You can also create movies using this version, thereby enabling you to share videos that show a user the information you feel is important. This version also allows you to import data and images from GIS systems your company might use (such as shapefiles from ESRI and MapInfo tab files), and provides the ability to batch geocode addresses, regionate large datasets, and automatically geo-locate GIS images. Using the Pro version with such datasets, you can take a dataset and quickly map thousands of addresses. You can also access demographic, parcel and traffic data layers, map multiple points on a map at once, and use tools that are not available in the free version.

**Table 1.2** Google Earth System Requirements

| Operating System | Minimum | Recommended |
|---|---|---|
| Windows XP or higher, with Windows 7 or 8 (or higher) recommended | ■ Pentium 3, 500 MHz CPU<br>■ 512 MB of memory (RAM)<br>■ 500 MB free hard disk space<br>■ Network speed of 128 Kbits/s<br>■ Graphics Card (Direct X9 and 3D capable with 64 MB of VRAM)<br>■ Screen resolution of 1024 × 768, 16-bit High Color with Direct X9 (to run in Direct X mode)<br><br>Outlook 2007 or higher is also required for email functionality | ■ Pentium 4.2 GHz or higher or AMD2400 xp or higher CPU<br>■ 1 GB or higher of memory (RAM)<br>■ 2 GB of free hard disk space<br>■ Network speed of 768 Kbits/s<br>■ Graphics Card (Direct X9 and 3D capable with 256 MB of VRAM)<br>■ Screen resolution of 1280 × 1024, 32-bit True Color |
| Mac OS X 10.6 or higher, with OS X 10.6.8 or higher recommended | ■ Any Intel Mac CPU<br>■ 512 MB of memory (RAM)<br>■ 500 MB free hard disk space<br>■ Network speed of 128 Kbits/s<br>■ Graphics Card (Direct X9 and 3D capable with 64 MB of VRAM)<br>■ Screen resolution of 1024 × 768, "Thousands of Colors" | ■ Dual Core Intel Mac CPU<br>■ 1 GB or higher of memory (RAM)<br>■ 2 GB of free hard disk space<br>■ Network speed of 768 Kbits/s<br>■ Graphics Card (Direct X9 and 3D capable with 256 MB of VRAM)<br>■ Screen resolution of 1280 × 1024, "Millions of Colors" |
| Linux (Kernel 2.4 or later, with 2.6 or higher recommended) | ■ Pentium 3, 500 MHz CPU<br>■ 512 MB of memory (RAM)<br>■ 500 MB free hard disk space<br>■ Network speed of 128 Kbits/s<br>■ Graphics Card (Direct X9 and 3D capable with 64 MB of VRAM)<br>■ Screen resolution of 1024 × 768, 16-bit High Color<br>■ glibc: 2.3.2 w/NPTL or later<br>■ XFree86-4.0 or x.org R6.7 or later | ■ 1 GB or higher of memory (RAM)<br>■ 2 GB of free hard disk space<br>■ Network speed of 768 Kbits/s<br>■ Graphics Card (Direct X9 and 3D capable with 256 MB of VRAM)<br>■ Screen resolution of 1280 × 1024, 32-bit color<br>■ glibc: 2.3.5 w/NPTL or later<br>■ x.org R6.7 or later |

While we will discuss these and other features in the chapters that follow, you will be left with the initial decision of what version to use. The Pro version does have a wide variety of capabilities that are useful for forensic investigations where you will work with location-based data. If you are planning on using images in reports and presentations that are distributed outside of your organization, the licensing agreement requires you to buy a license copy of the Pro version. While a free version might work better for your budget, take comfort in that it is a relatively inexpensive product that provides a great deal of benefits.

## INSTALLING GOOGLE EARTH ON YOUR COMPUTER

Now that you have a better understanding of what is available, let us go through the process of installing the Pro version on your computer. Once installed, you will be able to better follow what is discussed in chapters that follow, and use Google Earth's features and functionality yourself.

### System Requirements

Before you can install Google Earth, you need to have a computer capable of running it. As seen in Table 1.2, there are minimum requirements for running Google Earth on a PC, Mac or Linux machine, although you should try to install it on a machine with the recommended requirements to gain better performance. In addition to these specifications, if you are using a corporate computer, your network administrator may have pushed down restrictions on installing applications. For this reason, if you do not have an administrator account, you should contact your IT department to install GE for you.

A high-speed Internet connection is vital to using Google Earth. When you first open GE, the program will connect to Google's servers, accessing data and images that you have view and work with. If you do not have an Internet connection, the 3D viewer that is used to show you content will appear black, and if the connection is slow or has other issues, you may notice that images, maps or other graphical content may appear blurry.

### Obtaining a Copy of Google Earth

If you do not have a copy of Google Earth Pro to use with this book, do not worry. A copy of GE can be purchased from their website, or you can download a 7-day trial version of Google Earth Pro. The trial version gives you a better understanding of the features available in Google Earth. Once the trial period is over, you will need to decide whether to buy it or install the free version.

In a browser, http://www.google.com/earth/ and click on the Google Earth Pro link on the page.

**FIGURE 1.1** Setup Screen for Google Earth.

If you are buying a copy, click the *Buy Now* button, otherwise click the *Download Trial* button.

If you wish to download a trial of the software, fill out the form that appears, and then click the *Sign me up!* button. Once this is done, just click the link to download Google Earth Pro, and after reading the license agreement, click the *Agree and Download* button.

## Installing Google Earth Pro

Now that you have downloaded the setup for Google Earth Pro, you are ready to begin installing it. When prompted in your browser, click the *Run* button and follow the instructions that appear on the screen. The steps in the installation are as follows.

1. After starting the setup executable, you will see a screen indicating the Google Earth Pro is being downloaded and installed. When complete, the following screen will appear. Enter your email address in the *Username* field (Figure 1.1).
2. Enter your license key in *License Key* field. Note that if you applied for a 7-day trial, you will have received this in your email. If you have purchased a copy, then you would enter the username and key associated with that license.
3. Click on the *Enable automatic login* checkbox so that it appears checked. This will automatically log you in each time you open GE in the future.

4. If you want to enable Google Earth for anyone using your computer, click on the *Enable for all users on this machine* checkbox so it appears checked.
5. Click the *Log In* button

## Bibliography


[1] ABC News. Alleged 'Ninja Robber' Claims Trio Used Google Earth to Target Homes; 2014, July 12. Available from: ABC News: http://abcnews.go.com/US/alleged-ninja-robber-claims-trio-google-earth-target/story?id=24534235 [accessed 12.09.2014].

[2] CBC News First Nations Learn to Map Territories Using Google Earth; 2014, August 25. Available from CBC News: http://www.cbc.ca/news/aboriginal/first-nations-learn-to-map-territories-using-google-earth-1.2746110 [accessed 08.09.2014].

[3] Shawnee Police Department Shawnee Police Department Crime Maps; 2010. Available from Shawnee Kansas: http://gsh.cityofshawnee.org/WEB/PoliceCMS.nsf/c0019294e957d2c28525754a004b58b4/4b7c35995b121854862575e5004a6574?OpenDocument[accessed 08.09.2014].

[4] The Telegraph Google Earth Used By Thief To Pinpoint Buildings With Valuable Lead Roofs; 2009, March 15. Available from The Telegraph: http://www.telegraph.co.uk/news/uknews/4995293/Google-Earth-used-by-thief-to-pinpoint-buildings-with-valuable-lead-roofs.html [accessed 10.09.2014].

[5] U.S. Fish and Wildlife Service View Wetlands Data with a Keyhole Markup Language (KML) File; 2014, May 1. Available from U.S. Fish and Wildlife Service: http://www.fws.gov/wetlands/Data/Google-Earth.html [accessed 08.09.2014].

CHAPTER 2

# Using Google Earth

**INFORMATION IN THIS CHAPTER:**

- Using Google Earth
- The Google Earth UI
- Navigation
- Views
- Google Maps
- Tours
- Configuration

## USING GOOGLE EARTH

Now that you have a better understanding of what Google Earth (GE) is, it is time to get a better understanding of how it can be used. In this chapter, we will discuss the user interface, its features, and how to navigate to desired locations. As we will see, you can control what information appears on a screen by using different views, and by toggling various elements on and off. Because you may want GE to run a certain way when it starts up, we will also discuss some of the preferences and options you can configure to get Google Earth to work the way you want.

## THE GOOGLE EARTH UI

Google Earth provides an intuitive User Interface (UI) to view a virtual globe, maps, and geographic information, while still having quick control over performing more complex tasks. Even if you have never used GE before, you will find that it is easy to use once, especially after you have worked with it a short time.

As seen in Figure 2.1, the UI can be broken down into several distinct areas. Each of the panes and components are numbered on the figure, and correspond to a description that follows. While we will discuss them in greater



Google Earth Forensics

**FIGURE 2.1** Google Earth UI.

depth in the sections that follow, this overview should provide useful to you as an annotated reference throughout the book.

1. *Search panel*. As we will see in a section that follows, this panel is used to search for locations and to manage the results.
2. *Toolbar buttons*. These buttons will be detailed in a section that follows, and specific functions will be discussed.
3. *Sign in*. This function is used to sign into a Google account to share on Google+ or to email what is being viewed in Google Earth.
4. *3D viewer*. This is where the user views the globe and terrain.
5. *Navigation controls*. As we will see in a later section, these are used for zooming in and out, looking and moving around in the 3D viewer.
6. *Overview map*. This gives a different location perspective of what is being viewed in the 3D viewer. It can be toggled on and off by clicking on the View menu, and then clicking Overview Map.
7. *Scale legend*. This legend is used to measure distances and can be turned on an off by accessing the View menu and clicking Scale Legend.
8. *Layer panel*. As we will discuss later in this chapter, this panel is used to display or hide features appearing on maps (e.g., hiking trails, etc.) in Google Earth.
9. *Places panel*. This panel is used to find, save, organize and visit placemarks.
10. *Status bar*. This bar is used to view coordinate, elevation, imagery date and streaming status.

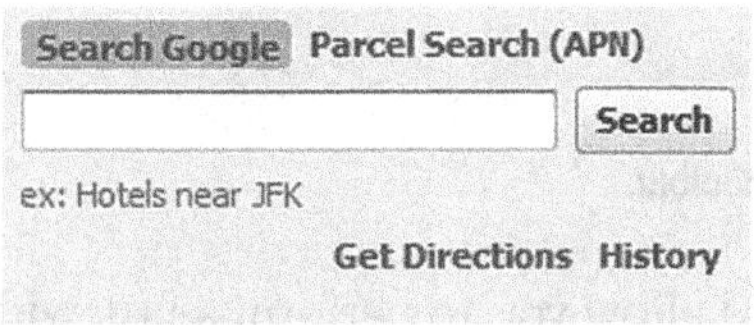

**FIGURE 2.2** Google Earth search panel.

## Search

One of the most used features in Google Earth is its ability to search for locations. As seen in Figure 2.2, the Search panel contains a field where you can enter the name of a point of interest, country, city, address, or coordinates. If you have selected the default mode to Search Google at the top of the panel, you will be searching the actual Google search engine for results. Because of this, you can use the same types of criteria that you would use when searching on Google's Web site, such as "Hotels near Las Vegas" to have a listing of hotels in and around Las Vegas appear in the 3D viewer. After entering the search criteria, you then click the Search button, and the 3D viewer will change to display that location.

As mentioned, you can also search by coordinates of a location. If you would like to try searching by coordinates, and do not know any off the top of your head, you can always search for a location (e.g., your city and state/province) in Google (www.google.com) using the keyword "coordinates". For example, if you entered "area 51 coordinates" into Google, it would return the coordinates 37.2350° N, 115.8111° W. By copying and pasting these coordinates into Google Earth, the 3D viewer would display an area view of Groom Lake and Homey Airport.

At the bottom of the Search panel, you will find two links named *Get Directions* and *History*. The *History* link allows you to view and manage past searches you have done. The *Get Directions* link is used to acquire directions from one location to another. By clicking this link, two fields will appear in the panel where you can enter the start and end locations. Upon clicking the *Get Directions* button, the 3D viewer will display a graphic of your route, and the Search panel will expand to display textual directions.

The other link at the top of the Search panel is *Parcel Search (APN)*. APN is an acronym for Assessor's Parcel Number, which is number associated with a parcel of land. If you had the property parcel number for a location, which could be obtained through government offices like a County Tax Assessor, you could search through U.S. parcel data to find that location.

## Toolbar

At the top of the Google Earth UI, you will find a toolbar like the one shown in Figure 2.3. Each of the buttons on the toolbar provide quick and easy access to

**FIGURE 2.3** Google Earth toolbar.

common functions, and allow you to perform a task with the click of a button rather than navigating to associated items in Google Earth's menus.

In later chapters, we will use a number of the functions on the toolbar. To provide a handy reference for you, the table that follows provides information on the individual buttons and their functions (Table 2.1). As you will see in the description column, there are associated menu items and keyboard shortcuts that can be used to perform the same actions.

## Layers

A layer is additional information that is overlaid on the map, satellite image or aerial image displayed in the 3D viewer. As seen in Figure 2.4, the Layers panel contains a listing of layers that can be turned on and off by, respectively, selecting or deselecting the checkbox beside it. For example, if you wanted to have the names of roads to on appear on an image in the 3D viewer, with lines clearly marking those roadways, you would click on the checkbox beside *Roads* so it appears checked. Similarly, if you did not want to see the names of places to appear on an image, you would click on the checked box beside *Places*, so it now becomes unchecked.

Some of the layers available in this panel are grouped together in folders, which can be expanded by clicking on the arrow to the left of an item. For example, by clicking on the arrow beside Earth Pro (US), you will see three layers that are only available in the Pro version of Google Earth. Checking these will display US Demographics, US Parcel Data, and US Daily Traffic Counts layers in the 3D viewer, providing you with additional information about an area.

The layers available can be extremely useful when reviewing geographic regions. For example, the *Borders and Labels* layer allows you to see international borders, the names of counties, and similar political boundaries. The *Places of Interest* layer will display various points of interest (POI), including businesses, airports, hospitals, schools, and even postal, city and school boundaries. When you enable this layer, you can use the information to find directions to and from that location, and see icons designating a POI that can be clicked to display additional information.

### *Viewing Current Conditions*

While many layers are used to add static content to an area being viewed, some layers connect to Google servers to display recent or current conditions. Because GE connects to the Internet and accesses information available through

**Table 2.1** Buttons on Google Earth Toolbar

| Button | Description |
|---|---|
| | *Conceal or display sidebar*. This shows/hides the left panes, providing a larger view of the 3D viewer. You can also hide it by clicking View \| Sidebar on the menu, or using the keys: Ctrl + Alt + B |
| | *Add placemark or location*. This allows you to add a placemark to a location that will appear in the Places pane of the UI. This allows you to save a location(s), which can later be used to create tours or snapshot views. You can also add a placemark by clicking Add \| Placemark on the menu, or using the keys: Ctrl + Shift + P |
| | *Add a polygon*. This allows you to add a polygon to a location that will appear in the Places pane of the UI, representing the area you selected. You can also add a polygon by clicking Add \| Polygon on the menu, or using the keys: Ctrl + Shift + G |
| | *Add a path or line*. This allows you to add a path or line to a location that will appear in the Places pane of the UI. Once a path is added, you can select it and play a tour of it. You can also add a path by clicking Add \| Path on the menu, or using the keys: Ctrl + Shift + T |
| | *Add an image overlay*. This adds an image over a map or terrain. You can also add a image by clicking Add \| Image Overlay on the menu, or using the keys: Ctrl + Shift + O |
| | *Record a tour*. This begins recording a tour, which you can then play and share with others. |
| | *Display historical imagery*. If previous satellite images and aerial photographs are available, you can click this button to display a slider on the screen, which allows you to move between current and older images taken on different dates. You can also access this by clicking View \| Historical on the menu |
| | *Display sunlight across landscape*. When clicked, a slider will appear on the screen, allowing you to move between sunlight different times of day. In doing so, you will see the 3D viewer grow lighter and darker. You can also access this by clicking View \| Sun on the menu |
| | *View sky, moon and planets*. When clicked a dropdown menu appears, allowing you to change between different content in Google Earth. Selecting Earth provides the default view of Earth data. Selecting Sky will show information on constellations, stars, and other objects in the sky. Mars will show information on the planet Mars and related missions to the red planet, while Moon will show content related to Earth's moon and lunar missions. You can also access this by clicking View \| Explore and then selecting the content previously mentioned. |
| | *Measure a distance or area*. This displays a dialog box allowing you to measure the length of a line or path, the circumference and area of a polygon or circle, and measure 3D buildings with a path or polygon. You can also access this by clicking Tools \| Ruler on the menu |
|  | *Email view or image*. Clicking this displays a dialog box asking you if you want to email a Placemark, View or Image of the content displayed in the 3D viewer. You can also access this by clicking File \| Email on the menu, and selecting Email Placemark (Ctrl + E), Email View (Ctrl + Alt + E) or Email Image (Ctrl + Shift + E) |
| | *Print current view*. Clicking this will print the current view in the 3D viewer to a printer. You can also access this by clicking File \| Print on the menu, or using the keys: Ctrl + P |
|  | *Show current view in Google Maps*. This launches the current view in Google Maps. You can also access this by clicking File \| View in Google Maps on the menu, or using the keys: Ctrl + Alt + M |

**FIGURE 2.4** Layers panel.

Google's servers, there are some layers that are able to display current or recent information about an area.

The Traffic layer allows you to see real-time traffic conditions in metropolitan areas. When you add this layer, GE retrieves real-time estimates of traffic conditions, which are overlaid on the map being viewed. If you were to zoom in on a highway or street with this layer activated, you would see traffic marked in green, yellow and red to respectively indicate that traffic is moving at normal, below normal, and slow speeds. By clicking on one of these colored areas, you can view more detailed information about the conditions for that section of roadway.

The Weather layer is used to display information about the current weather conditions of an area being viewed. By expanding this folder, you can choose to view sublayers of information related to current and forecasted whether, cloud patterns, and radar images of rain, snow and ice conditions.

## Places

The Places panel is used to save locations that you have visited, and navigate to ones that have previously been stored. As seen in Figure 2.5, the panel contains folders and sites that have been added, as well as placemarks you have created.

There are several ways you can add places to this panel. After searching for a location, you can right click on a location in the Search panel, and click *Save to My Places* on the context menu that appears. In doing so, the location will be stored under the My Places folder in the Places panel. These locations will be identifiable in the panel by the red icon beside the entry's name.

Another method of adding places is through the use of placemarks. Earlier in this chapter, we discussed how the toolbar has a button to add a placemark or

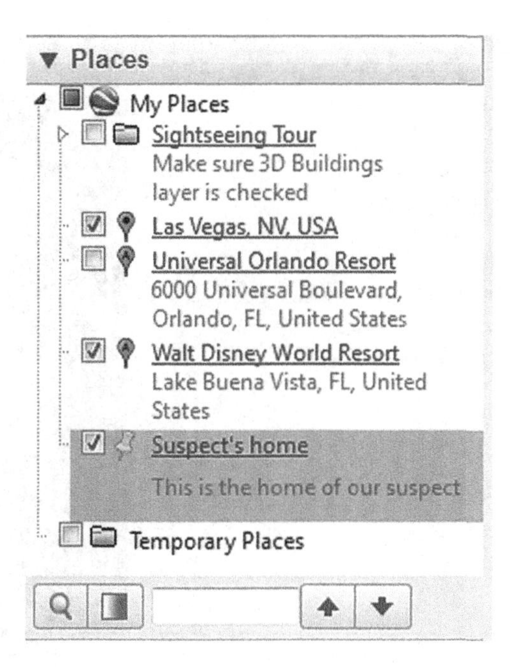


**FIGURE 2.5** Places panel.

location. After navigating to the location you want to save in the 3D viewer, click on the Placemark button on the toolbar. An icon that looks like a pushpin will appear on the screen, which you will position over the location you want to placemark. Upon clicking your mouse, the *New Placemark* dialog box shown below will appear (Figure 2.6).

The *New Placemark* dialog box allows you to details about the location. Fields and tabs on this dialog include:

- *Name*, which is where you would enter a meaningful title for the placemark.
- *Latitude* and *Longitude* fields, which inform you of the location of the placemark.
- *Description* tab, where you can enter text, links and images.
- *Style, Color*, which allows you to choose a color, scale and opacity for the placemark icon and label.
- *View*, where you can modify the latitude, longitude, range, heading tiles and date/time. If you want to simply apply the current view in the 3D viewer (i.e., the same altitude and angle) to this placemark, click on the *Snapshot current view* button on this tab.
- *Altitude*, which allows you to select the height that the placemark appears over the terrain. You can enter a numeric value, or use the slider bar on this tab to adjust the setting.

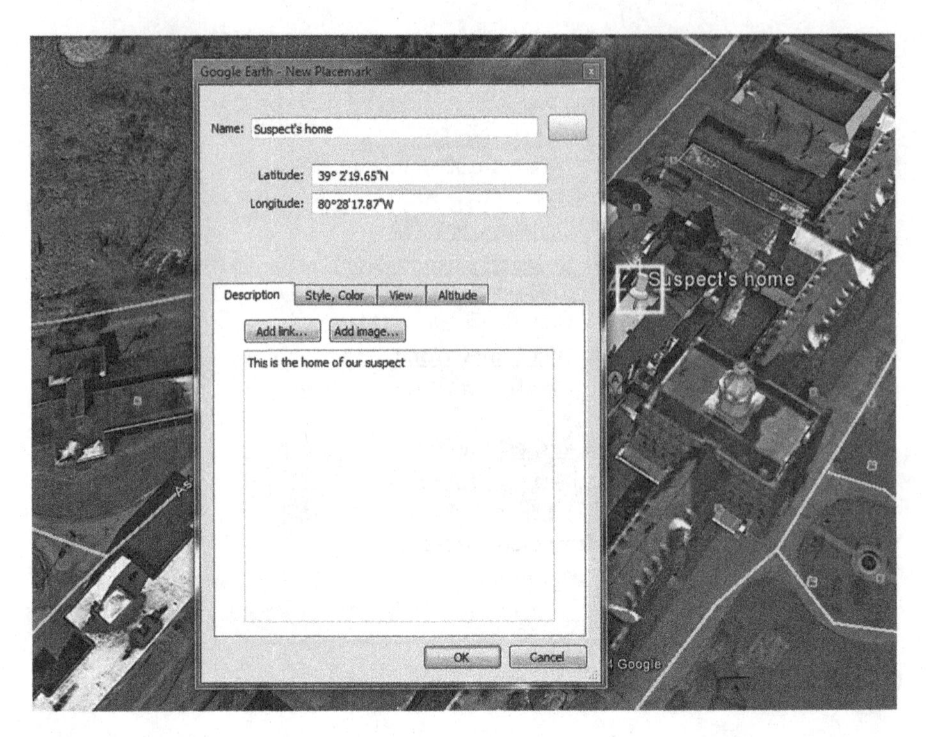


**FIGURE 2.6** Creating a new placemark.

In the top right of the dialog box, you will also notice a button with an icon. Clicking this, you can choose a different icon to make your placemark appear unique from others you might add. There is a large selection of alternative icons to choose from, and you can even add a custom icon, or choose no icon if you wish. After clicking the OK button, the placemark is added to the Places panel, and you will see that it has the same icon beside the text. This helps to make it stand out from locations you have searched for and added to the panel for future reference.

If you have a lengthy list of items added to the Places panel, it can be difficult browsing through them to find what you want. At the bottom of the pane, you will notice a search field that can be opened by pressing CTRL-F on your keyboard. By typing in the name of a placemark or location you have added and clicking the magnifying glass button to the left, Google Earth will go to the place that either matches or partially matches the text you are searching for.

---

Latitude and longitude allow you to pinpoint any location on Earth. Latitude shows you how far north or south you are located relative to the equator and longitude indicates how far to the east or west you are located relative to the prime meridian – a line drawn through Greenwich, UK.

Latitudes to the north are progressively larger positive numbers while latitudes to the south are progressively larger negative numbers. So therefore when you see a positive latitude you know that it is north of the equator and conversely when you see a latitude that is negative it is south of the equator. The higher the positive or negative number the further north or south it is.

This same scheme holds true for latitudes - negative numbers are the distance from the prime meridian to the west and positive numbers are the distance from the prime meridian to the east.

Latitudes and longitudes are commonly shown in one of the following forms.

- *Sexagesimal degrees*. This is an older system that uses degrees, minutes and seconds as its measured.
- *Decimal degrees*. This is the modern system makes the minutes and seconds into a fraction of a degree.

## NAVIGATION

The 3D viewer is where most of your navigation takes place. Google Earth provides multiple ways of moving around the 3D viewer, inclusive to mouse movements, keyboard shortcuts, and navigation tools located in the upper right hand corner.

### Mouse Controls

The mouse can be used to interact with the 3D viewer in several ways. When you move the pointer over the 3D viewer it changes into an open hand. By clicking the left mouse button, it will appear to make a grabbing motion, changing from an open hand to a closed one. When this happens, you can then drag in any direction to change your point of view.

You can also use the mouse to cause a drift over the earth, making the 3D viewer continue moving in a particular direction after you have released the mouse button. This effect is useful when recording tours, which we will discuss later in this chapter. To drift, click and hold the mouse briefly then move in a direction and release the mouse. Imagine this movement as if you were tossing something in the direction you move the mouse. To stop the motion, click in the 3D viewer. The drift will also come to a halt on its own if left alone.

#### *Zooming*

To use the mouse to zoom into a location, double click and Google Earth will begin to zoom into the point clicked. The zoom can be halted by single clicking, and started again by double clicking once more. If your mouse has a scroll wheel, you can also use that to zoom in and out. Holding the ALT button in combination with using the scroll wheel causes the zoom to go in smaller increments.

Another way to zoom in is to position the cursor on the screen and press the RIGHT mouse button (OPTION on a Mac). Once the cursor changes to a double arrow and a crosshairs appear, move the mouse backward or pull toward

**FIGURE 2.7** Tilting the view.

you, releasing the button when you reach the desired elevation. Your view will always zoom to the crosshairs. Zooming out is accomplished with the same steps, but by moving the mouse in the opposite direction.

### *Tilting View*

When zooming in, you will notice that the display in the 3D viewer will tilt as it approaches ground level. It automatically adjusts from a downward view to one comparable to standing on the ground. If your view was at a higher elevation (i.e., above ground level), you can manually tilt the view to shift from a top-down view to one that is more level with the terrain. For example, as seen in Figure 2.7, tilting the view over the Grand Canyon allows you to shift the aerial view to one that is more parallel with the horizon. In doing so, you get a different perspective of the location, gaining a better understanding of the terrain.

You can manually tilt the view by pressing the middle button or scroll wheel on a mouse, and then moving the mouse forward or backward. This can also be done by holding the shift key and scrolling with the scroll wheel. If you do not use a mouse with a scroll wheel, you can still tilt the display. Simply, hold the left mouse button and Shift button on your keyboard as you drag backward or forward. When tilting is activated the crosshairs appear, and you will notice the perspective in the 3D viewer change.

### *Looking*

Google Earth also provides the ability to look around without moving in any direction, mimicking a head turn. To look around from a fixed vantage point, hold the CTRL key (COMMAND on Mac) and the left mouse button and drag.

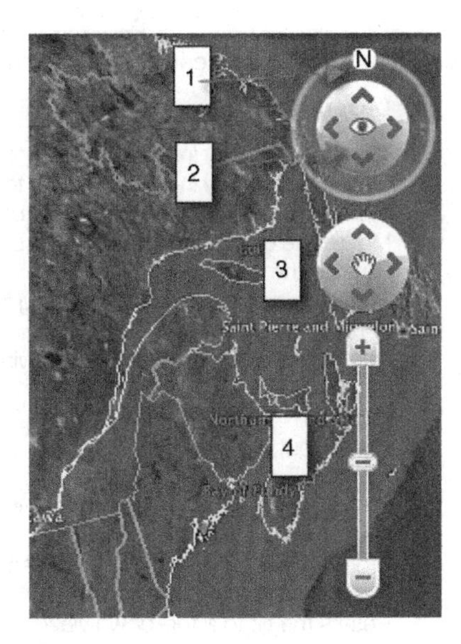


**FIGURE 2.8** Navigation tools.

## Navigation Controls

The navigation controls provide the same functionality for viewing the Earth as the mouse controls. As annotated in Figure 2.8, each set of controls allow you to move, tilt

1. The top control allows you to rotate the view. When the outer ring is clicked or dragged, the view in the 3D viewer rotates. Clicking the **N** button on the ring will reset the view so that North is at the top of the screen.
2. Within the ring is a circular control with arrows, which is used to rotate the view as if you were turning your head and looking around. This tool is called the *Look Joystick*. Like the ring round it, this can be clicked and dragged to change the direction of the view.
3. The *Move Joystick* is another control works similar to a joystick, allowing you to move around the map. You can click the arrows or click, hold and drag to smoothly change the location.
4. The slider control at the bottom is used to zoom in and out. As you approach ground level, tilting occurs to bring your view parallel to the ground. As we will discuss later in this chapter, the tilting feature can be turned off or on by configuring the navigation options

As you work with the navigation controls, you will notice they are set to automatically fade away when not in use or the focus of the mouse pointer. If for some reason the navigation controls do not appear in the upper right hand corner when the cursor is placed near them, adjust the setting by clicking the

**Table 2.2** Keyboard Shortcuts to Navigate in Google Earth

| Keys | Description |
|---|---|
| Left arrow | Moves the display in the 3D viewer to the left |
| Right arrow | Moves the display in the 3D viewer to the right |
| Up arrow | Moves the display in the 3D viewer upward |
| Down arrow | Moves the display in the 3D viewer downward |
| Shift + right arrow | Rotates the view clockwise |
| Shift + up arrow | Tilts the view up, so you are viewing from a top-down perspective |
| Shift + down arrow | Tilts the view down, so you are viewing from the perspective of a horizon |
| Ctrl + up arrow, Pg Up | Zooms in |
| Ctrl + down arrow, Pg Down | Zooms out |
| Enter | Zoom into a selected placemark or item that you've selected in the Places panel |
| Spacebar | Stop the current motion or movement in the 3D viewer |
| N | Resets the view so that north is up |
| U | Resets the tilt to a top-down view |

*View* menu, selecting *Show Navigation* and clicking either *Automatically* (to have it fade in and out) or *Always* (to have it constantly displayed). Alternatively, you could also select the *Never* menu item so it does not appear at all, or *Compass Only* to simply display a compass in the right hand of the 3D viewer.

### Keyboard Shortcuts

As we have seen throughout this chapter, there are a number of keystrokes that can be used to do specific actions in Google Earth. Once you have become familiar with these shortcuts, pressing a combination of keys can be quicker and easier, especially for tasks that are common and repetitive. As seen in Table 2.2, there are a number of keys that can be used for navigating to and around locations.

While these are limited to actions related to navigating in Google Earth, there are of course numerous other keystrokes that can be used to perform specific tasks and access menus. As we have done throughout this chapter, we will continue to mention shortcuts you can take with your keyboard to perform certain actions.

## VIEWS

There are numerous ways to view information through the 3D viewer. When you first start Google Earth, you are presented with a virtual globe, which you can zoom in on to explore images from satellites and aerial photos. Using various layers, you can toggle additional information and see details of the terrain, and access facts about a particular area to gain a greater understanding of vicinities. If you zoom in as far as you can, you can even view photos and representations of what an area looks like on the ground.

### 3D Views

As you navigate to various locations, you will notice that not all of the places depicted are actual photos. Google Earth has created 3D representations of buildings, landmarks, and other natural and artificial features found in those sites. The objects are textured and realistic, and provide a good understanding of what a city or other locale looks like as you zoom in and explore the area.

The 3D objects in Google Earth are created by Google as well as third parties. 3D modeling programs like SketchUp (www.sketchup.com) are used to create models, which are then imported into GE. While earlier versions of this program were owned by Google and free to use, it is now proprietary software that is currently owned by Trimble (www.trimble.com), although there is a free version that is available for non-commercial use.

The 3D renderings of buildings and other objects can be turned on and off using the Layers panel in Google Earth. As we discussed earlier in this chapter, if the checkbox beside a layer is checked, then it will appear in the 3D viewer.

### Ground View

Until now, we have focused on how Google Earth provides an aerial view of the world, but it also allows you to see things at ground level. By zooming in as far as possible to a particular area, the view will tilt and level out, showing a rendering of what it is like to stand at that spot. This view uses various layers and 3D models to extrapolate what the area would look like from the ground. Once the view has changed to ground level, you can use the navigation controls to move and look around.

An alternate method of switching to ground view involves the icon of a peg-like person standing on a circle, which is shown in Figure 2.9. The icon will appear below the navigation controls when you zoom in on a location to an altitude of approximately 500 km or less. By clicking on the icon and dragging

**FIGURE 2.9** Pegman icon.

it onto an area you are interested in, you will switch to a ground-level view of that location.

While the 3D buildings and Terrain are turned on by default in the Layers panel, you will want to ensure these are enabled if you are going to use the ground-level view. Missing this content from the 3D viewer will take away the enhanced realism of the view, and take away models that may be important to what is viewed in an area.

## Street View

If you have used Google Maps in the past, you are probably already familiar with Street View. While the Ground View will show you 3D buildings, trees and other objects, Street View shows actual photos that were taken of an area. In Google Earth, you can enter street view by dragging the pegman icon onto a street that has a blue border around it. As soon as you do, you are provided with street level images, allowing you to see the location as if you were standing on the street.

Once you are in either Ground View or Street View mode, you can navigate and rotate the view of a location by using the arrow keys on your keyboard, the scroll wheel of your mouse, double clicking on areas, or dragging the image to move left, right, or (in some cases) upward. If you drop the pegman icon onto an area where Street View is not available, Google Earth will instead provide you with a Ground View of location.

As seen in Figure 2.10, you can switch between these two views with the click of a button. By clicking the button with the pegman icon on it, the 3D viewer will display street level photographic images, while clicking on the building icon will display 3D renderings of the location in Ground View. The button to the right can also be used, allowing you to exit whichever of the two modes you are currently in.

## Viewing Historical Imagery

As mentioned earlier in this chapter, Google Earth not only allows you to view current images of an area, but also historical aerial photos from the past. This feature is particularly noteworthy if you needed to view how an area looked in the past, such as when working a cold case that occurred years ago in a particular location.

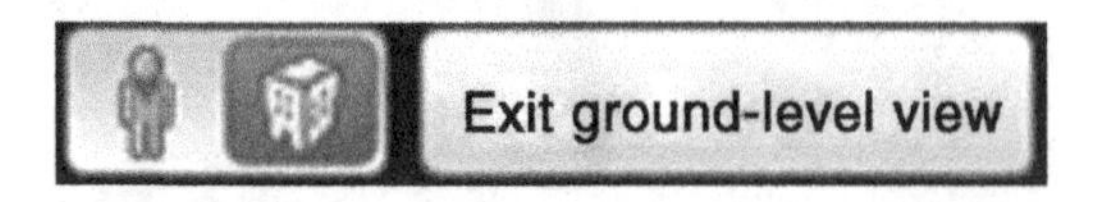


**FIGURE 2.10** Toggle between ground and street views.

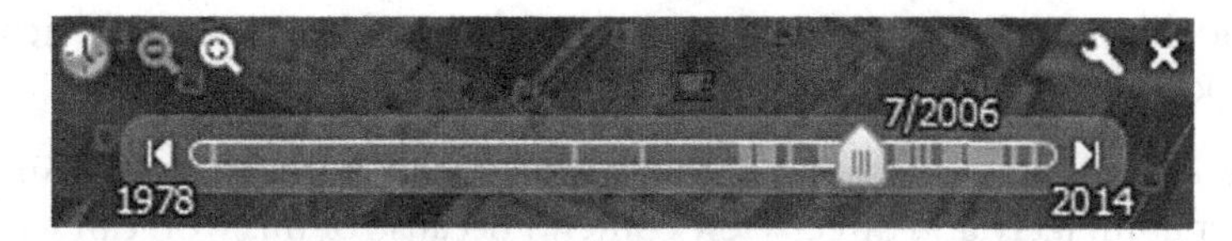


**FIGURE 2.11** Slider control for viewing historical imagery.

When older images of an area are available, you can click the Clock button on the toolbar to display a slider bar, similar to what is shown in Figure 2.11. By moving this slider bar to the left, or clicking the left arrow button on the slider, the imagery in the 3D viewer will change to display images from previous dates. Small vertical lines on the slider bar provide a timeline, indicating when the aerial images were taken. To move forward in time, you would move the slider to the right, or click the right arrow button beside the slider. For example, to view the most recent imagery available, you would move the slider to the rightmost point on the bar.

When viewing historical imagery, you should consider turning off 3D buildings and other layers that may obscure how a location actually appeared in the past. For example, if you were to look at the location of the World Trade Center in New York City, you would see a 3D representation of One World Trade Center (Freedom Tower) appearing at the site. Without turning off 3D buildings, this model would remain there as you viewed imagery in the past, prior to its construction. The model not only obscures how the area looked in the past (appearing overtop of actual aerial photos), but also provides an inaccurate representation of the area at various points in history.

## View in Google Maps

Google Maps (http://maps.google.com) is another product from Google that we have mentioned. Like Google Earth, it allows you to view satellite imagery, aerial photos, maps, and has features like Street View, tours and navigation that we discuss in this chapter. It also provides features that allow you to plan routes for traveling by car, foot, bicycle, or public transit. Unlike Google Earth, it is not a standalone program, and does not need to be installed on a computer. You can access Google Maps using a browser or mobile app.

One of the advantages Google Maps has over Google Earth (at the time of this writing) is that it provides historical Street View imagery, while Google Earth only allows you to view historical imagery from above. Using the Street View feature in Google Maps, if historical imagery is available, you will see a clock icon in the upper left of the screen. By clicking this, you can view older images of the same location.

While there are similarities, you should not assume that one is a watered down version of the other. Google Earth has many tools and features that Google

Maps does not, including the ability to view historical aerial imagery, rich data layers, powerful tools, and additional 3D content.

Google Earth is integrated with Google Maps, with both products using the same aerial imagery and Street View content. Because of this, you can view the same location seen in Google Earth in Google Maps. To view what is displayed in Google Earth in Google Maps, simply click on the *File* menu and click *View in Google Maps* (CTRL + ALT + M).

## TOURS

Tours can be created to show maps, locations and other information in a way that guides a person through the information you want to convey. Using a tour, you can record yourself navigating in the 3D viewer, and optionally add audio to explain what is being shown. The content that is created can be compelling, and showcase important points about what is being shown.

### Tour Guide

To see what a tour looks like, you can view ones created by others, which are available in Google Earth. At the bottom of the 3D viewer, you will see a strip of tours similar to what is shown in Figure 2.12. These tours are presented in a feature called the *Tour Guide*, and can be played by simply clicking on the one that interests you. The tours available will change as you navigate to different locations, so you are given the opportunity to view content that is related to areas you are interested in.

At the top left of the Tour Guide, you will see a downward arrow that can be clicked to minimize it in the 3D viewer. Once minimized, it will appear as a small button in the bottom left of the viewer, with an upward arrow. Clicking this will restore the Tour Guide to its previous full size. If you do not want the Tour Guide to appear at all in Google Earth, you can show and hide it by clicking on the *View* menu, and then clicking *Tour Guide*.

When you click on a tour, it will begin to play and a small control will appear in the lower left-hand corner of the 3D viewer, providing you with buttons to go back, play/pause, and fast forward. These buttons are similar to what you would find on a DVD player. There is also a slider bar that allows you to move back and forth through the tour, and a replay button to watch it again.

**FIGURE 2.12** Tour guide.

**FIGURE 2.13** Controls for recording a tour.

## Recording Tours

While it may be interesting to view the tours others have created, Google Earth also allows you to create and narrate your own tours. In doing so, you can explain the important points related to a case, so that others can view it later.

To record a tour, you would begin by clicking the *Record a Tour* button on the toolbar, which we discussed earlier. As soon as this is pressed, a control will appear in the lower left-hand corner of the 3D viewer, as shown in Figure 2.13. By pressing the Record button (shown with a dark gray circle), any actions or movements in the viewer will begin to be recorded. You will see that it is recording because the button will turn red, and the counter to the right will begin to change, indicating the duration of your tour. To record audio, you would click on the button with the microphone icon, which will turn blue while audio is being recorded with the tour. Clicking either of these buttons a second time will stop recording.

To see how this works, let us use some of the things we have learned in this chapter, and record a short tour by following these steps:

1. In the *Search* panel, type the address of your workplace and then click the Search button shaped like a magnifying glass. When your address appears below this field, right click on it, and then click *Save to My Places*.
2. Repeat step one, but this time search for your home address, and save it to My Places.
3. Click the *Record a Tour* button on the toolbar.
4. When the control appears, click the *Record* button.
5. Use the slider control on the right side of the 3D viewer to zoom in as far as possible.
6. When the display changes to Street View, use the controls on the right side of the 3D viewer to rotate and position yourself in front of your house.
7. In the *Places* panel, select your work address and press the Enter key. The 3D viewer will change, flying out to show the location of where you work.
8. Press the *Record* button a second time to stop recording.

Once you have stopped recording, you will see the control shown in Figure 2.14 appear in the lower left-hand corner of the screen. To the left of this control,

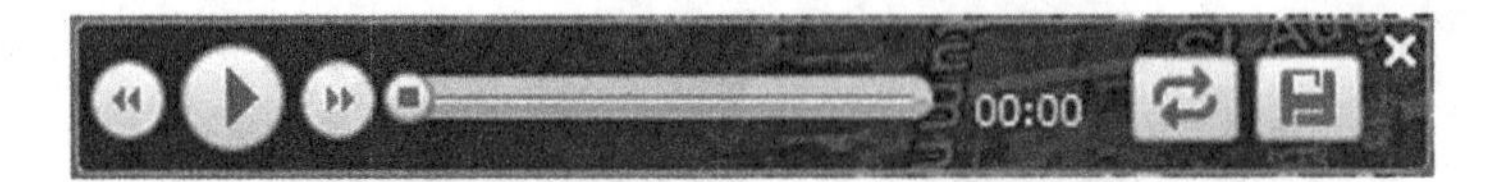


**FIGURE 2.14** Controls for recording a tour.

you will see buttons to go back, play/pause, and fast forward. The control also provides a slider to move through the tour, and information on the minutes and seconds you have progressed into it. To the right of the timer, you will see a button that allows you to replay the footage, and another to save it.

When you click the *Save* button, a dialog box will appear, allowing you to name the tour, and provide a description that can include images and links. After clicking the *OK* button, the video is saved to the Places panel in Google Earth. If you wanted to save this to a file that could be shared with others, you can right click on the tour you just saved in Places, and then click *Save Place As*... In doing so, a dialog box will appear asking where you'd like to save it.

Google Earth allows you to save a tour as either a KMZ or KML file, which we will discuss thoroughly in Chapter 4. These files can then be loaded into Google Earth by others, so they can then watch the tour you have created. In Chapter 6, we will expand on your knowledge of creating tours, and show you how you can directly apply this to an investigation.

## CONFIGURATION

When you first use Google Earth, there are a number of default settings that can be modified at any time. To access these settings, you would click on the *Tools* menu, and then click the *Options* menu item. Once this is done, a dialog box with the following tabs will appear:

- *3D View*, which is used to control settings used by the 3D viewer.
- *Cache*, which allows you to control resources used by your computer.
- *Touring*, which provides settings for recording and playing tours. We will discuss this tab in detail in Chapter 6.
- *Navigation*, which sets elements related to navigating in the 3D viewer.
- *General*, which allows you to set other preferences related to the program.

### 3D View

Using the *3D View* tab, shown in the following tab (Figure 2.15), you can set preferences for how Google Earth displays things in the 3D viewer. In changing these settings, it is important to realize that as a general rule adding enhancements to the 3D view will require more system resources from your computer

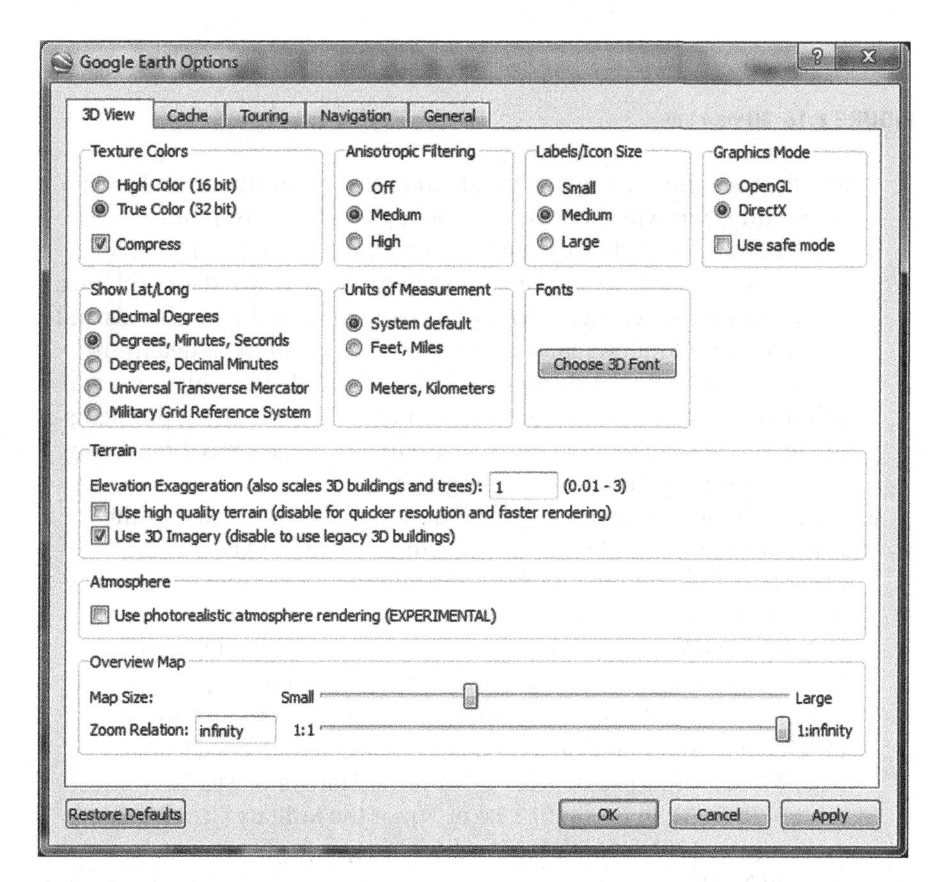


**FIGURE 2.15** 3D view tab.

and may affect performance. As such, you may want to test how your system works with a setting before deciding whether to leave any changes permanent.

As with the other tabs we will discuss, the various options are grouped together into related sections. Those on the 3D View tab include:

- *Texture Colors*. This feature is used to set the number of bits for color representation. The higher the value the more realistic the view.
- *Anisotropic Filtering*. This filters pixels in texture mapping that produces a smoother looking image. When this feature is enabled it produces a much smoother image around the horizon when the Earth is viewed from a tilted angle. This is a graphics card memory dependent feature and should only be used if your graphics card has at least 32 MB of memory. It is set to off by default.
- *Labels/Icon Size*. This feature affects the default size for labels and icons in the viewer. Use the small setting if you are working with urban areas

Imagery Date: 10/8/2012 32° 17.165' N 110° 48.418' W elev 2592 ft eye alt 20.17 mi

**FIGURE 2.16** 3D view tab.

and the medium for a mix of urban and more expansive areas. Larger icons and labels will show better from higher elevations in the viewer.

- *Graphics Mode.* Google Earth either uses either OpenGL or Direct X rendering on Windows machines and only OpenGL for Mac and Linux. The program will let you choose what setting is best for your needs and hardware. On Windows machines Google Earth will try to determine what settings appear to be optimal for your installed graphics card and will make suggestions. There is a safe mode to trouble shoot problems in the 3D viewer. Enabling safe mode will turn off advanced rendering features reducing the load on your graphics card.
- *Show Lat/Long.* When you move the mouse across the surface of the Earth, the latitude and longitude coordinates are displayed in the lower left of the viewer. The default display of the coordinates is in degrees, minutes, seconds (DD.MM.SS) or degrees, decimal minutes (DD.MM.MMM). As seen in Figure 2.16, these coordinates appear beside other information about the imagery your mouse point is hovering over, inclusive to elevation and altitude. You can choose the Degrees option to display geo-coordinates in degrees decimal (e.g., 37.421927°-122.085110°), Universal Transverse Mercator (e.g., 580954.57 m E 4142073.74 m N), or the Military Grid Reference System (e.g., 17SND457761127) which is a geographic reference system used by NATO.
- *Units of Measurement.* Google Earth uses the system default unless specifically told to use the Imperial or Metric system of measurement. The display will change from feet to miles when appropriate and from meters to kilometers.
- *Fonts.* Selecting the Choose 3D Font button brings up the font selector that allows you to choose which font will be used for most text and labels.
- *Terrain.* This section allows for the display of terrain - higher quality will affect performance of the program and possibly the system. Google Earth allows for "Elevation Exaggeration" to adjust the appearance of hills and mountains. The value can be adjusted from zero to three with decimal values allowed. The default setting for this is one.
- *Atmosphere.* This setting is currently labeled "Experimental" and allows for rendering of the atmosphere. Note that enabling this feature may cause some performance hits depending on the machine you are using.
- *Overview Map.* The overview map is used to determine position in relation to the entire Earth. The default setting for this is 1:infinity

and means that regardless what is being currently displayed in the 3D viewer the overview map with show the entire Earth indicating general position with a red square or crosshair. When the map zoom is adjusted to 1:*n* where "*n*" is equal to the number the user sets multiplied by the current view allowing you to effectively to zoom out of the current view by whatever factor you choose.

## Cache

The *Cache* tab is used to optimize performance on the system used to run Google Earth. The tab provides a field to set the *Memory Cache Size* to control how many megabytes (MB) of memory is used by GE, and is limited by how much physical memory is installed on the machine. You can clear out what is stored in memory by clicking the *Clear memory cache* button on this tab.

The disk cache refers to space on your hard disk, which is used to store imagery on the machine and view it offline. If you were not connected to the Internet, GE would try to load previously stored images from the hard disk. The *Disk Cache Size* field controls how much data can be stored on the hard disk, but is limited to 2 GB. You can clear out the cache by clicking the *Clear disk cache* button on this tab.

If you have a slow or intermittent Internet connection, and have sufficient disk space, you may want to consider increasing the size of the disk cache. The more space Google Earth has to store imagery locally, the less it will need to access from the Internet. This will provide you with smoother zooming, and allow layers to load faster.

## Navigation

The *Navigation* tab is used to set preferences that control how you move around in Google Earth, and how the application itself will operate when taking you from one location to another. As shown in Figure 2.17, the tab provides the following groups of options.

- *Fly To Speed*. A slider allows you to set how fast or slow you zoom in and out of locations. Alternatively, you can also adjust the speed by typing a new value into a field beside it. Adjusting this to a lower speed will give you more time to discuss what is happening when you are recording a tour, and need to explain the transition from one location to another.
- *Mouse Wheel*. If your mouse has a scroll wheel, you probably use it to zoom in and out of locations. The slider in this section allows you to control how fast this occurs, while the checkbox allows you to invert the direction of zooming using the wheel
- *Non-mouse Controller*. If you use a joystick or another type of controller, other than a mouse, you can use options in this section to enable the

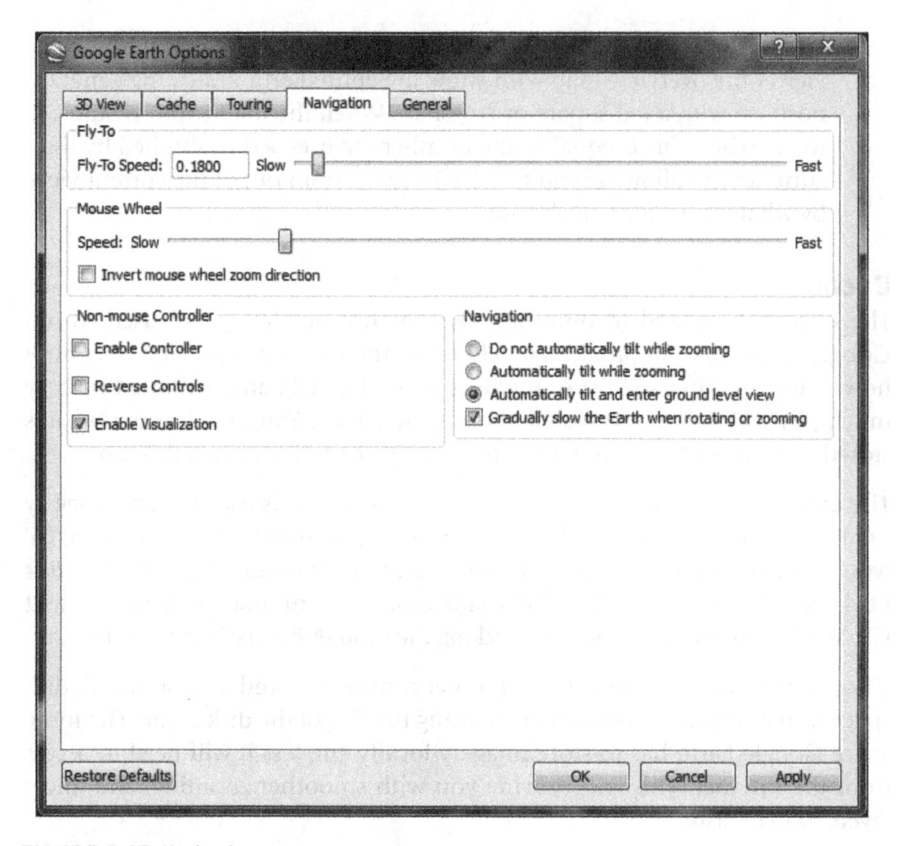


**FIGURE 2.17** Navigation tab.

controller, reverse controls on the device, and control whether arrows appear on the screen.

- *Navigation*. The first two options in this group controls whether the perspective will tilt in the 3D viewer as you zoom into a location. The third option allows you to automatically tilt and enter Ground view. The final checkbox in this section allows you to set whether gradually slow and stop when you spin or zoom in on the Earth. If this is not set, spinning the globe (such as when it appears at startup) will have the globe continually spin indefinitely, and will lack the gradual decrease in speed.

## General

As seen in Figure 2.18, the *General* tab provides a wide variety of settings relating to how Google Earth will function. The various groups of options on this tab are as follows.

- *Display*, which sets whether tool tips and building highlights are displayed, and if Web results can be viewed in your browser.

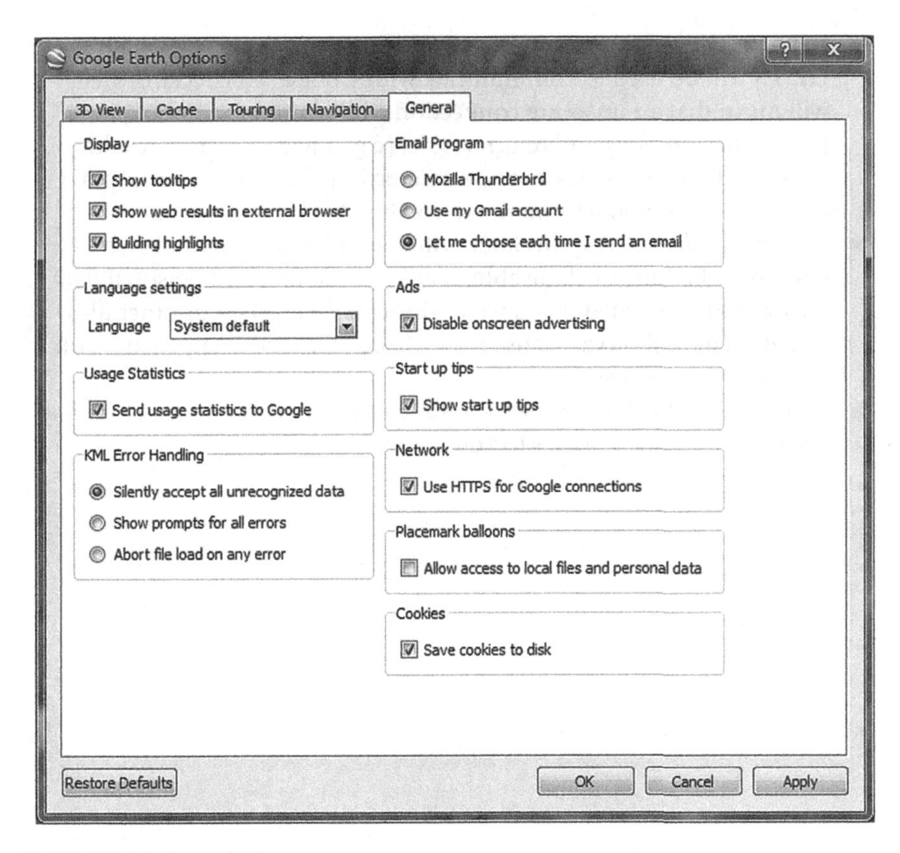


**FIGURE 2.18** General tab.

- *Language Settings*, where you can set the language used.
- *Usage Statistics*, which determine whether information about your use of the program are sent to Google. When using this application for forensic purposes, it is suggested that you uncheck this checkbox, as you probably want to limit the information being sent out to Google.
- *KML Error Handling*. Options in this section determine what Google Earth will do when it encounters a KML file containing errors. In Chapter 4 we will discuss KML in great detail, and see how these files are used to store geographic features and modeling information for use with Google Earth.
- *Email Program*. This section is used to specify which email application will be used to send data to people you specify. In Google Earth, you can click on the File menu, click Email, and then select whether to email a placemark, view or image to others.
- *Ads*. The checkbox allows you to disable onscreen advertising
- *Startup tips*. When you first start Google Earth, you are presented with tips that can help you learn how to better use the application. To turn off this feature, uncheck the box in this section.

- *Network*. The checkbox in this section allows you to specify whether HTTPS will be used to communicate with Google. Unchecking this will mean that an unsecure connection will be used to send requests (e.g., search criteria) and return results (e.g., imagery, search results, etc.). As HTTPS provides a secure connection between your computer and the servers, you will want to have this checkbox checked.
- *Placemark balloons*. The checkbox will determine whether links on placemark balloons will be able to access local files and personal data. This would be needed if you were adding links to PDFs or other files stored on a local server, or used a KML file that used images stored on your local hard drive.
- *Cookies*. This checkbox allows you to control whether Google Earth has permission to save cookies to your computer.

CHAPTER 3

# GPS, GIS, and Google Earth

**INFORMATION IN THIS CHAPTER:**

- Understanding GPS (Global Positioning System)
- Understanding GIS (Geographic Information System)
- Geo-location Information in Pictures

## UNDERSTANDING GPS

It is almost to an odd sight to someone unfolding a map in their car, and trying to navigate to a destination using the lines and images printed on the large sheet of paper. If you are like most people, you probably use a GPS system to find your way from point A to point B. Listening to audio cues and visual instructions, people will follow the directions provided by navigation systems mounted in the dashboard of their cars, apps on smartphones, and hand held devices. Navigating to a destination has never been easier.

Because so many people use GPS units, it is not surprising that they can be a useful source of evidence. Using these units, people travel to *waypoints*, which are sets of coordinates that identify a physical location, and if you look in the history of locations you have visited using a GPS navigation unit, you will notice that these coordinates have been saved to the device. From a consumer perspective, it is useful to see the recent locations so you can set the GPS unit to take you there again. For an investigator, you should consider how this information might be useful to see where someone has gone, and when they went there.

### What is GPS?

Even if you own a GPS device or used one for years, you may be unaware of how GPS works, and how it came to be. GPS is an acronym for the *Global Positioning System*, and was developed by the US Military to overcome previous



Google Earth Forensics

limitations in navigation systems. Originally called the Defense Navigation Satellite System (DNSS) and later Navstar-GPS, the GPS project was launched in 1973 and merged ideas from several other projects that were developed during the Cold War Arms race.

In 1983, Korean Air lines Flight 007 was shot down by the USSR for straying into its prohibited airspace. This lead to a directive by then President Ronald Reagan to make GPS freely available for civilian applications once the system was sufficiently developed. The first satellite in the current constellation was launched in 1989 and the last – the 24th – was launched in 1994.

The United States government owns and operates GPS as a national resource and the US Department of Defense is the custodian of the system. Starting from 2000 the U.S. has made improvements to the system including new signals and increased accuracy for military, civilian and commercial needs.

Several rivals to the U.S. GPS include the Russian GLONASS (which has a full complement of 24 satellites for global coverage), the European Galileo (begun in 2011 and expected to be completed in 2019 with 30 satellites), the Chinese BeiDou Navigation Satellite System (operational in 2011 and offering services to Asia-Pacific area with 10 satellites and expected to be global in 2020) and the Indian Regional Navigational Satellite System (the first satellite to be launched in 2013 and completed in 2014 providing service only to India).

## Components

While GPS is generally referred to as a single entity, it is actually comprised of a separate systems working together. The GPS system is composed of three major groupings:

- Space segment (SS)
- Control segment (CS)
- User segment (US)

### *Space Segment*

The GPS system was designed to have at least 24 satellites functioning at any given time. There are a number of spares in the case of failure and the total number of healthy and in orbit satellites is 32 as of December 2012. Satellites are also known as space vehicles or SV.

The satellite constellation consists of six orbital planes consisting of four satellites each and are centered on the Earth and fixed in relation to the stars. The orbits of the satellites is arranged so that there at least nine in the line of site anywhere on the earth and insuring that the four minimum required for position in each plane are unevenly spaced in the orbit. This gives the system considerable redundancy in case of satellite failure.

The satellites orbit at an approximate altitude of 12,600 miles (20,200 km) with a radius of around 16,500 miles (26,600 km). Individual SVs make one complete orbit each 11 h 58 min and a total of two per day. The time period for the two orbits of each SV is called a sidereal day which is a time scale based on the Earth's rate of rotation measured relative to the fixed stars.

### *Control Segment*

The control segment is further divided into four subsections, consisting of a master control station (MCS), an alternate MCS, four ground antennas, and six monitor stations. Amongst other things, the control segment is responsible for monitoring the flight paths of the SVs to synchronize atomic clocks, adjust the ephemeris (see the later discussion of error correction) of the satellite's internal orbital model and perform navigational updates using inputs from the monitoring stations, space weather information and others.

### *User Segment*

The user segment consists of the many military, scientific commercial and civilian users of GPS and their many receivers.

## Trilateration

To get a precise location with GPS the satellites use something called trilateration. We will explain basic 2D trilateration using an example.

Say that you are participating in a scavenger hunt. You find a clue to the next piece that you need to collect. It tells you that the location of the next piece is in a coffee shop, that is 5 miles from the hardware store, 10 miles from the library and 8 miles from the sandwich stop. How would you locate it?

First you would draw a circle centering on the hardware store that had a radius of 5 miles. Then you would draw a circle centering on the library that had a 10-mile radius. You might notice that you now have two circles that intersect at two points. One of these two points is the location of your next clue.

To get the precise location you now draw a third circle that centers on the sandwich shop with a radius of 8 miles. You now see that this third circle intersects with the others on one of the two points giving you the exact location which is in the middle of the Williams Park (Figure 3.1).

To get trilateration in three dimensions it is a little trickier but the general concept is the same. Imagine that now instead of single radius extending out from the center point many radii of the same length extending our to form a sphere.

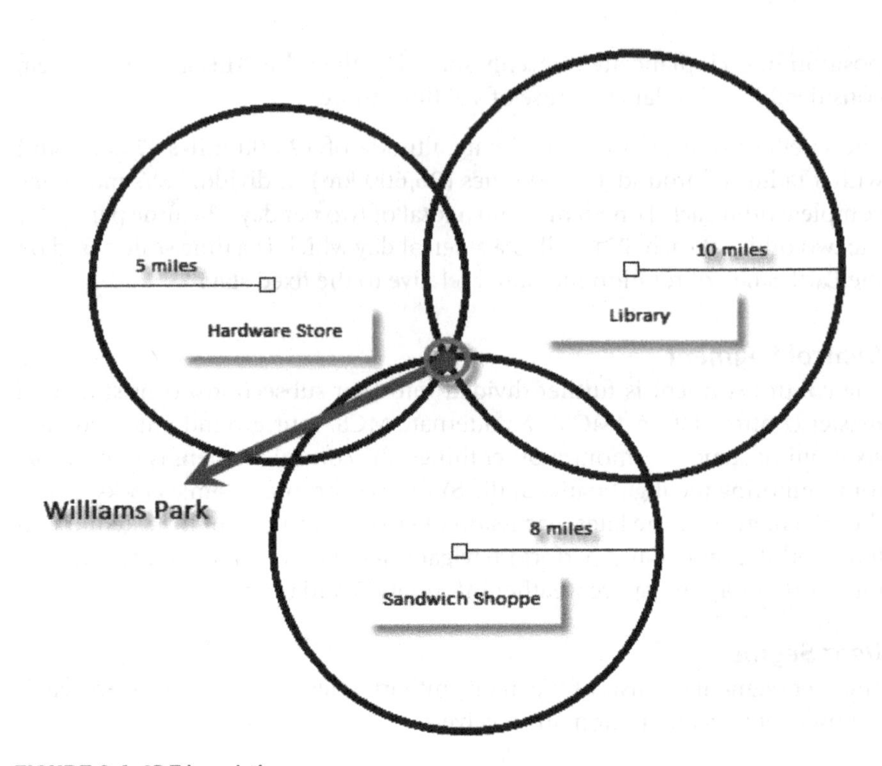


**FIGURE 3.1** 2D Triangulation.

In our new thought experiments each of our center points now represents a satellite – let us call them A, B and C.

Our distance from Satellite A is 20 miles. We can postulate a sphere from the center of this satellite. If we are also 25 miles from Satellite B then we can also imagine a sphere from it that overlaps Satellite A. These spheres will intersect in a perfect circle.

If we now take our distance from Satellite C – 15 miles – and draw out a sphere from there, it will intersect the circle formed by A and B in two points.

Now the Earth is also a sphere and one of these two points is located on its surface, the other in space. The GPS receiver will reject the point in space, leaving the other point as the location in question. In general, receivers will poll from at least four satellites to get more accurate location and altitude details.

So, for trilateration to occur we need the following information:

1. At a minimum, three satellites in view
2. The distance between the point you are located at and the orbiting satellites in view.

How we calculate the distance between your receiver and the satellites is the subject of our next section in our GPS 101 primer – measuring distance.

## Measuring Distance

GPS receivers calculate distance from satellites by deciphering high frequency, low power radio signals – as a pseudo code – that are broadcast from the satellites. Since radio signals travel at the speed of light in a vacuum – 300,000 km/s (186,000 miles/s ), receivers calculate the distance to the satellite by timing how long the signal takes to get to them. In theory this is accomplished by a simple mathematical expression: Distance = Velocity × Time.

---

Pseudo Random Code is a complicated digital code – pulses of on and off. It is so complicated that it looks like random radio noise, hence the term pseudo random code.

---

However, there is a wrinkle in the equation. In order for the proper time lag from the satellite to the receiver to be measured accurately, the clock of the satellite and the clock of the receiver need to be in perfect sync. This is a problem because satellites use super precise atomic clocks which are very expensive – costing between 50,000 and 100,000 dollars – which makes them impossible to use in consumer products.

There is a solution to the atomic clock problem. The receiver uses a quartz clock that it constantly resets by checking its own inaccuracy against the timing of four or more satellites. The time value of all the satellites will cause all the signals being received to align in a single point in space and this therefore is the correct time that the receiver will set its clock against. This makes the receiver clock almost as accurate as the satellites atomic one.

Finally, the receiver needs to know where the satellites are for the distance information to be of use in positioning. Since satellites travel in high and predictable orbits, this is not too hard. Satellites broadcast something called an almanac that the receiver stores that tells it where a satellite is supposed to be. There are some things that need to be corrected for, atmosphere, effects of the sun and moon. These are called ephemeris errors because they affect the satellites orbit or ephemeris. The Department of Defense continually monitors the ephemeris errors using precise radar and transmits updates to the satellites so they can be used in the satellites broadcasts.

---

Time to first fix – TTFF – is a measure of the time required for a receiver to acquire satellite signals navigation data, and calculate a position. This is commonly broken down into three types

*Cold*. This occurs when the receiver is missing or does not have accurate information on satellites. The receiver must first locate a satellite and then download the almanac to get other information on other satellites. The almanac is broadcast every 12.5 min.

*Warm*. The receiver knows the current time within 20 s, a current position within 100 km, and a velocity within 25 m/s, and it has valid almanac data. It now must obtain ephemeris data - broadcast every 30 s and valid for 4 h.

*Hot*. The receiver has all the ephemeris, almanac, time and position data and can thereby rapidly obtain signals from orbiting satellites. The time that a receiver takes to calculate a new position is called the Time to Subsequent Fix or TTSF.

## Error Correction

The timing and distance calculations we described in the previous section work well - in a vacuum. However, there are things that can cause errors in the calculations of the receiver.

GPS signals need to travel through the Earth's atmosphere in order to get to the receiver. There are two layers that the signal passes through that need to be accounted for - the Ionosphere and the Troposphere.

The ionosphere is the layer of the atmosphere that ranges from 50 km to 500 km above the surface of the Earth. This layer consists mostly of charged particles that interfere with the signals from satellites. The errors from the ionosphere are larger accounted for by mathematical modeling - though it remains the most significant source of errors in GPS.

The other layer of the atmosphere that influences GPS signals is the troposphere. This is the lower layer of the atmosphere that contains Earth's weather. Water vapor fills this layer and it is variable both in temperature and in air pressure. It causes error but not as significant as the ionosphere. Errors from this layer can be corrected with modeling as well as other techniques.

Another type of error that can occur in GPS is something called multipath error. Signals can also bounce off of natural as well as manmade objects on their journey to the ground. This also can be corrected. We look at solutions for error correction in our next section

## Error Solutions

Most of the errors caused by the Earth's atmosphere can be removed by the aforementioned modeling. For instance, mathematical models can be used to account for the charged particles of the ionosphere and the variable contents of the troposphere. The receiver must factor in each angle of signals coming in from satellites as it is the angle that determines the length of the trip of the signal through the atmosphere.

Something called Dual Frequency measurements can also be used but it is very sophisticated and requires an advanced receiver as well as the military having the only access to one of the frequencies required for this type of calculation.

Fortunately civilian companies have developed – albeit complicated – solutions to this issue.

Multipath error can be corrected by the receiver accepting only the earliest arriving signals instead of the later ones – it is assumed that the earlier ones are the direct signals.

## Differential GPS

Differential GPS (DGPS) is a type of error correction that compares the mobile receiver with a stationary receiver at a known location. Because the differential station knows its location, it can handily calculate a receiver's inaccuracy. The station will broadcast a radio signal to all the DGPS equipped receivers in the area providing signal correction.

Smartphones or devices use a variant of GPS that is called Assisted GPS or A-GPS. The development of A-GPS was accelerated in the United States due to the Federal Communications Commission's requirement that the emergency system be provided with a cell phone's location.

This 1996 ruling by the FCC required wireless carriers to implement technology that determines and transmits the location of someone calling 911, so that emergency services would know where a person was calling for help. Although this was vital for enhanced 911 services, the goal of A-GPS is to speed up the Time to First Fix of a device by providing the ephemeris data through the cellular network. This data is then used in conjunction with the GPS chip in the device to provide location and navigation.

The A-GPS system is also used alongside wifi positioning and cell-site multilateration. A detailed discussion of A-GPS is beyond the scope of this course and the student is encouraged to avail himself of the list of resources to further research this and related concepts.

## Importing GPS Data into Google Earth

At this point, we have provided you with an understanding of how GPS works, and how information on your location is acquired for such devices as tablets, mobile phones, apps, navigation systems, and so on. The data related to your physical location may be stored as data on a device, which can then be retrieved through various tools, and analyzed to determine the current and previous locations a person visited. In Chapter 5, we will discuss a number of tools that can be used for the acquisition of such data, but you can acquire and analyze data using Google Earth.

Google Earth (GE) allows you to import data from GPS devices through a direct connection or by importing files acquired from a device. Once this is loaded into GE, you can see the waypoints stored in the device, and review the locations through the 3D viewer. Doing so provides an easy way to see where the coordinates correspond to specific locations on a map or aerial view. In Chapter 6 (as part of a case scenario we will work through) we go step-by-step through the process of importing GPS data into Google Earth. However, while we detail these steps later, a basic understanding of the process is given here.

- For a direct connection, you would attach a Garmin or Magellan GPS unit to your computer using a USB or serial port connection, and (after turning on the device) click on the *Tools* menu and click *GPS*. The screens that follow will connect to the unit, and import its data into Google Earth.
- For GPS files, you can click on the *File* menu, click *Open*, select a GPS file and follow the import process.

Once GPS data has been added to Google Earth, it appears in the Places panel. You can save the GPS data as a KML file to the hard disk. As we will see in Chapter 4, a KML file is a format that is used to store information on geographic features and modeling. Once the file is saved in this format, others using applications that support KML files (such as Google Earth or GIS software discussed later in this chapter) can load and view it. To save the data in GE, simply right click (CTRL-click on a Mac) on the GPS folder in Places and then click *Save Place As*.

Because GPS devices record the date and time associated with a particular waypoint, you can navigate through the imported data using the date and time. In doing so, you can see where a person was when they carried a mobile GPS unit, and the date and time they visited that location. Once you have imported GPS data into GE, Time Controls appear in the upper left-hand corner of the 3D viewer. A slider on this control allows you to move between locations recorded by the GPS unit. Sliding the control the rightmost point will display the most recent waypoint recorded, while sliding the control to the left will show you places they visited earlier. You can also view this as an animated series of events in Google Earth. To view it as an animation, select the GPS data folder in the Places panel. On the Time controls that appear in the top left of the 3D viewer, click the *Play* button.

## UNDERSTANDING GIS

GIS is an acronym for *Geographic Information Systems*, and refers to applications and hardware that is used to capture, store, manipulate, analyze and manage spatial or geographic data. The spatial data used by these systems may be acquired from a wide variety of sources, and geo-referenced to locations around the world. In doing so, the information allows you to identify the location of natural and constructed features, boundaries, water sources, and so on. With such data, a GIS system could be used to create detailed maps that display information on maps, satellite images and aerial photography.

Many organizations use GIS systems for a variety of uses where geographic information needs to be displayed and analyzed. Some of the examples of how it is used by law enforcement include creating maps of crime scenes for use in investigations and court, displaying calls for service made with Computer-Assisted Dispatch (CAD) systems, and crime mapping. Because police and other law enforcement agencies have data on where crimes have occurred and officers have been dispatched, Crime Analysts can use GIS to plot the locations on a map. In doing so, they can identify hot spots where certain crimes have occurred, as well as trends and patterns in certain locations.

By overlaying datasets showing demographics, the locations of certain points of interest, and other data, Crime Analysts can get a better understanding of the causes of crime. For example, if vandalism is rampant near a particular school, you might consider that the students of that school may be involved. Similarly, if hate crimes occurred near the hangout of a street gang or organization that dislikes a group being victimized, it shows a possible underlying cause of the crimes are occurring in that area. This information is useful in solving crimes, and also for creating strategies for deploying officers. For example, if pawnshops or convenience stores were being held up in part of the city, the mapped information would show the need for a greater police presence. Seeing where other stores are located in that area would also show other potential targets of crime, and indicate the need for more patrols in those neighborhoods.

## GIS Data and Google Earth

Google Earth Pro allows you to import GIS data directly, so it becomes part of what is displayed in the 3D viewer. This can be a powerful feature for investigations using GE. For example, if you were looking at the locations a suspect visited, you might have loaded that person's GPS data into Google Earth. You could then overlay GIS data on crimes committed in that area and instantly see a correlation of where the suspect was, and crimes occurring at those locations. Once this is done, you will be able to save the map so others can view it in GE, include it in a tour of crime scenes, or print it for use in court.

When you add data from a GIS system into Google Earth, there are two types of data that can be imported.

- *Vector*, which are points, lines or polygons that appear on a map and represent various objects. For example, these may be used to indicate a road or lake at a location. Vector datasets that would be imported into Google Earth would include ESRI shapefiles or MapInfo tab files.
- *Raster*, which are grids of data representing images, such as those from a satellite or aerial photographs, continuous surfaces (e.g., elevation models), or thematic classes (e.g., land cover, habitat maps, etc.). Raster datasets you would import would include GeoTIFF files.

When importing the GIS data, you can expect there to be numerous files. For example, if you use ESRI's ArcGIS, then you would be importing the shapefile with the extension .shp, which is the core file containing feature geometry, an index file (.shx), and a .dbf file that contains attributes for each shape. Because you will want the data to appear in its correct position in Google Earth, it is important that you include files that have the correct coordinate system defined. For example, if you were importing shapefiles, you will need to include a projection file (.pri) associated with the data.

### *Importing Vector Data into Google Earth Pro*

To import vector data into Google Earth Pro, you would do the following:

1. On the *File* menu, click *Import*
2. When the dialog box opens, select the data type you are importing from the *Files of type* dropdown list. For example, if you were importing shapefiles, you would select *ESRI Shape (*.shp)*, while those importing MapInfo files would select *MapInfo (*.tab)*
3. Select the file being imported and click the *Open* button
4. If you are importing a large dataset, it may effect performance, so a message will inform you of the number of features contained in the file, and offer you the option of importing all of them or a selection of features. Click *Import all* to import the entire dataset.
5. A message will appear asking if you want to apply a style template for the features. If you click the *Yes* button, you will be presented with a *Style Template Settings* dialog where you can set the colors, labels and icons. Clicking *No* will skip this option.

Once the GIS data has been imported, it will be appear in the Places panel under the *Temporary Places* folder. If you want it to be available the next time you open Google Earth, you should drag the file out of this folder so it appears under the *My Places* folder. To have information from the GIS dataset appear in the 3D viewer, you would click the checkbox beside the name of the file you have just imported. To hide this information, ensure the checkbox is unchecked.

### *Importing Raster Data into Google Earth Pro*

Importing a raster dataset into Google Earth Pro is similar, but does have a few different steps. For example, if you were to import a GeoTIFF, you would do the following:

1. On the *File* menu, click *Import*
2. When the dialog box opens, select the data type you are importing from the *Files of type* dropdown list. In this case, we would select *GeoTIFF (*.tif)*
3. Select the file being imported and click the *Open* button

Because the raster dataset is geo-referenced to coordinate systems, the image overlay you have just imported will appear in the 3D viewer in the correct location. It will not need to be moved or adjusted. You will also see that the GeoTIFF now appears in the Places panel with the name you gave it. As with the vector dataset, you would ensure the checkbox beside this name is checked to have it appear in the 3D viewer, or uncheck it to hide the image overlay.

### Converting GIS Data into KML

The ability to import GIS data into Google Earth is limited to the Pro version. If you are using the free version, then you will need to convert this data into a KML file, which can then be imported into the free version of Google Earth. If you are sharing a dataset that is already been is imported into Google Earth Pro, you can save it as a KML file in the same way that we saved GPS data that was imported earlier in this chapter; simply right click (CTRL-click on a Mac) on the GIS data in Places and then click *Save Place As*.

GIS systems will generally provide tools or features to convert them. For example, in ESRI ArcMap 10 have the ability to convert a layer to a KML file. There are also products available to do the conversion, but some may require proprietary software to run. An example of this is the ArcGIS ArcScript found at http://arcscripts.esri.com.

## GEO-LOCATION INFORMATION IN PICTURES

In the twenty-first century, it is not too often that you come across people taking photos with film. Most people use their smartphones, tablets, or digital cameras to take a photo, and these photos commonly conform to the *Exchangeable image file format* (Exif). Exif is a standard that specifies image formats, inclusive to tags used by digital cameras (inclusive to the ones on your mobile devices). Some of the metadata tags used by Exif store such information as:

- Data and time that a picture was taken
- Camera settings that were used to take the photo (e.g., aperture, shutter speed, etc.)
- A thumbnail to preview the photo on the camera's screen
- Descriptions
- Copyright information
- Geo-location information

It is this last tag that is of particular interest to us. Since many cameras and mobile devices have a built-in GPS, it is able to determine its current longitude and latitude, and sometimes even it is the current altitude of its location. When a picture is taken, the camera can then add this geographic information to the

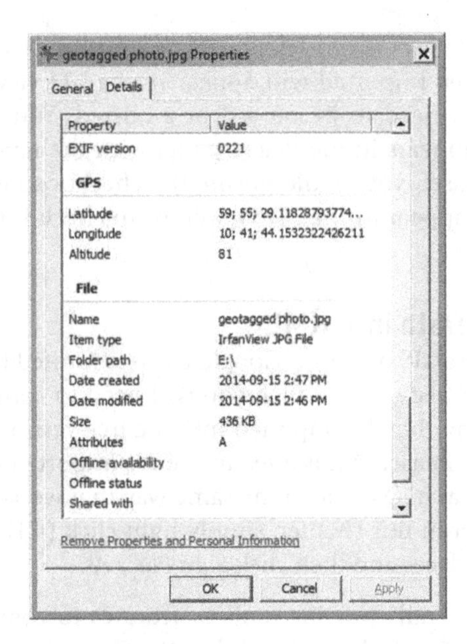


**FIGURE 3.2** Digital image properties.

tag in the photo. This process of adding geographic coordinates to a digital photo is called *geotagging*.

Viewing the geotagged information on a digital image is simple by doing the following.

1. Once the image is saved, navigate to its location using Windows Explorer
2. Right click on the image, and when the context menu appears, click Properties
3. When the Properties dialog appears, click on the Details tab
4. Scroll down to the section called GPS, and review that section

As seen in Figure 3.2, if the image has GPS information stored in a tag, you will be able to see the Latitude, Longitude, and Altitude associated with the image. As we saw in Chapter 2, you can then enter these coordinates into the Search panel of Google Earth to find the location where the photo was taken.

The metadata showing location information may also be added to other digital media, including video, Web sites, SMS messages, RSS feeds and other data found on Internet servers, network servers, computers and other devices. The data can be used to search for location-based news, sites and media, but (as we have seen) can also be used to identify the location of where an event occurred or where a suspect was located.

In addition to this information being automatically added at the time a photo was taken, it can also be manually geotagged by associating a photo to a mapped location, typing the coordinates into an HTML page, or using tools. Many sites support the process of geocoding and reverse geocoding media for use with certain features. *Geocoding* is a process of taking non-coordinate based information like a person's street address, and then determining the coordinates for that location, while reverse geocoding resolves coordinates to a location. Photosharing sites like Flickr (www.flickr.com) and social bookmarking sites like del.icio.us (https://delicious.com) support the addition of metadata to photos so others can search for them by location. Panoramio (www.panoramio.com) also supports geotagging, but as it is owned by Google, it also allows you to upload photos that can then appear as a layer in Google Maps and Google Earth.

CHAPTER 4

# KML/XML/HTML

## INFORMATION IN THIS CHAPTER:

- Markup Languages and Google Earth
- XML
- KML
- HTML

## MARKUP LANGUAGES AND GOOGLE EARTH

Google Earth (GE) allows you to display and modify geographic information and other content using markup languages. *Markup languages* are used by applications to control what is presented, influencing the information you see. These languages may be used to describe content and how it is to appear (such as the formatting of text or graphics), or describe how certain data is to be used by an application. In Google Earth, markup languages are used to create and modify files that determine what is shown on the screen, inclusive to map locations, placemarks, points, 3D models, images, and other data shown in the 3D viewer.

Because the code written with these languages can have a powerful effect on what appears in Google Earth, it is important to understand how they are used in the application, and how modifications to the code can benefit you. As we progress through this chapter, you will find that learning them may help you in the future with using Google Earth and creating your forensic evidence presentations.

### Different Kinds of Markup Languages

Although a wide variety of markup languages have been developed over the years for different purposes, only a few are used by Google Earth. Three of the markup languages we will discuss in this chapter are:

- HTML (Hypertext Markup Language), which was developed in 1989 by Tim Berners-Lee for controlling the presentation of information in Internet-based documents (i.e., Web pages).
- XML (Extensible Markup Language), which was developed by the World Wide Web Consortium (W3C) as a way to store structured data. It is used to describe the data stored in an XML document. As we will discuss later in this chapter, many applications are designed to read XML files.
- KML (Keystone Markup Language), which is used by Google Earth and other 2D and 3D mapping applications to display geographic data.

## A Basic Example of Using a Markup Language

To illustrate some of the basic ways a markup language works, let us look at one of the easier languages to learn, and look at a simple Web page written in HTML:

```
<html>
<body>
<p>This <em>is</em> the body of an <strong>HTML</strong>
document.</p>
</body>
</html>
```

In looking at the HTML above, you will notice that there are areas of text encased in angle brackets. These are called *tags* which are used to describe different content, and instruct an application how to handle content between the opening and closing tag. When an Internet browser loads a page containing this HTML, it will read the tags when it processes the page and use the information to determine how content is to be displayed. It will do this using the following logic:

- Text between <html> and </html> describes the whole document, indicating that everything between the opening and closing tag is an HTML document
- Text between <body> and </body> describes the body of the document, which will appear in the browser screen
- Text between <p> and </p> describes a paragraph, indicating that everything between the tags is a single paragraph
- Text between <strong> and </strong> describes that the text should appear bolded
- Text between <em> and </em> describes that the text should be emphasized, and appear italicized

If you were to copy the previous HTML code into a basic text editor like Notepad, and save it as a file with the extension .html, you could load it as a Web page in your Internet browser. In doing so, you wouldn't see anything but a line that reads "This *is* the body of an *HTML* document." The tags themselves are hidden. The tags are instructions are used by the application (in this case a Web browser), while you only see the result of it reading those instructions.

As we go through various markup languages in this chapter, you will find that there are many similarities in the basic structure of how they are written. Each starts by defining what language is used, so an application can understand how to process it. For example, an HTML document will start with the <html> tag, while a file written in XML (which we discuss later) will have the word *xml* as the first word in the opening tag. They will also use opening and closing tags with content nested between them. These closing tags are noticeable in that they use the same word or letter as the opening tag but start with a forward slash ("/").

## USING HTML IN GOOGLE EARTH

HTML is useful in improving how your forensic presentations appear in Google Earth. Using HTML, you can do such things as format text, insert hyperlinks to open Web pages and other documents, and insert graphics. For example, in Chapter 2 we showed you how you can add a placemark to a location in the 3D viewer, and enter a description of that location on the Description tab. As seen in Figure 4.1, text can be added to this box, providing information about

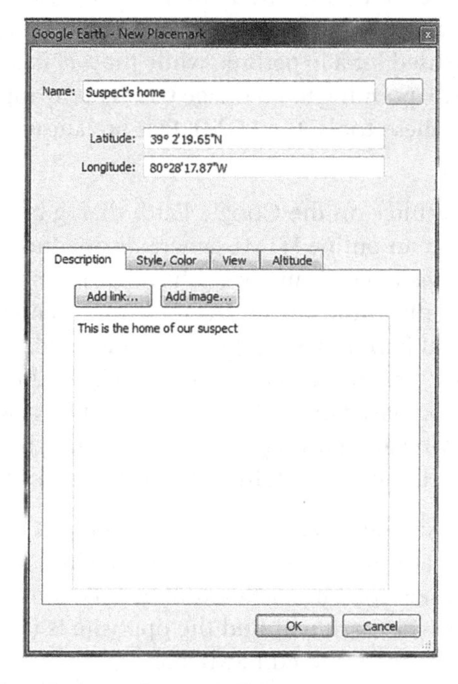


**FIGURE 4.1** Description tab of new placemark dialog.

the location. However, if you wanted to enhance the description, you could use HTML tags to format the text or add additional content.

When adding HTML to the description box, you do not need to worry about adding <html> or <body> tags, since (for reasons discussed later in this chapter), Google Earth can recognize when HTML tags are being used in the text you type here. For example, if you wanted to bold the words "our suspect", you would modify the text in this box to read <strong>our suspect</strong>. Similarly, if you wanted to add a link to some related Web page, you could enter the HTML code for a hyperlink to appear. For example, if we were going to add a link to Google, we would type the following in the description box:

```
<a href="http://www.google.com">Google</a>
```

In doing so, the <a> tag defines a hyperlink is being created, while *href* specifies the URL that is to be used. In this case, we have said the value of the URL is the Google Web site, which appears in quotes. The word "Google" that appears between the opening tag (<a>) and closing tag (</a>) is a word that will appear to people as a hyperlink. When people click this word, the hyperlink will be used to take them to the associated site or page.

If all this seems overwhelming, try not to worry. To make it easier, the text box on the Description tab of a New Placemark provides easy-to-use tools to add links and images. As seen in Figure 4.1, an *Add link…* button can be used to provide a URL to be used for a hyperlink, while the *Add image…* button allows you to specify the URL pointing to an image you want to appear in the description. When you use these tools, the HTML tags are automatically entered for you.

In addition to the features on the Google Earth dialog box, you can also use development tools, or an online HTML generator or editor to create the code. For example, if you visit HTML.am (www.html.am), you will find an HTML Generator. As seen in the following screenshot, you can enter text, and then use buttons similar to that found in a word processor to bold, italicize, or perform other tasks to format text and display data. By clicking the *Source* button, the content of the text you have entered and formatted will change to display the HTML code. By copying and pasting this into the dialog box in Google Earth, you have added HTML content with little effort (Figure 4.2).

As you look at a markup language, you will find that you become more familiar with its elements, attributes, and syntax. After a while, it can become just as easy to write or edit raw code in a text editor as it is to use a tool. That being said, there are times when you will find the opposite is true, as when you are using Google Earth to create and edit KML files.

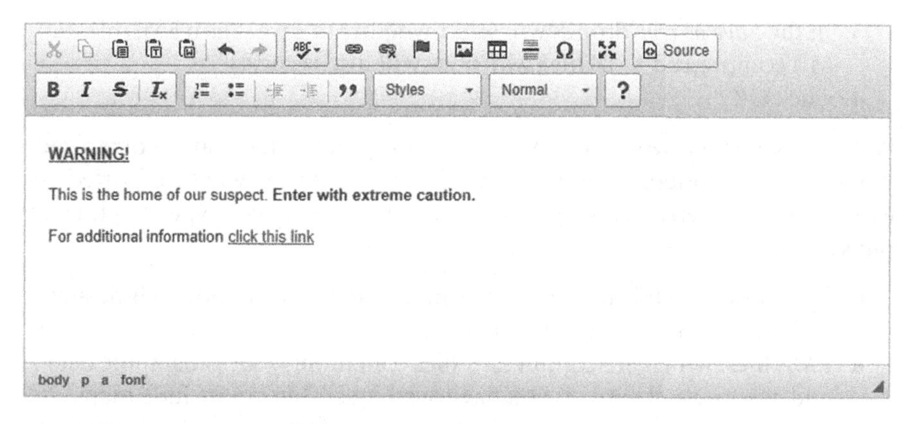


**FIGURE 4.2** Online HTML generator at HTML.am.

## WHAT IS KML?

KML is an acronym for the *Keystone Markup Language*, and used to display geographic data in Google Earth. Its name comes from the original name for Google Earth, which was Keyhole Earth Viewer, a product created by Keyhole Inc. The application experienced a name change after it was acquired by Google in 2004. Even though KML was developed for use with Google Earth, it became an international standard of the Open Geospatial Consortium in 2008.

Google Earth uses KML to depict the geographical features related to a location in GE's 3D viewer, inclusive to points, lines, images, polygons, and models. Data describing the geographic overlays, coordinates, textures, camera angles and other content related to a location can be stored in a file, and used by Google Earth to determine what should be displayed. Data in the KML file can be modified using features in GE, or by using the KML markup language to directly edit the code.

### Saving KML Files

In the same way that HTML is a markup language that is used to present content through a browser and can be saved to a file, the same holds true for KML files. While an HTML file is displayed in an Internet browser, KML files are displayed in Earth browsers like Google Earth. In addition to GE, other applications can also display KML, including Google Maps, ESRI ArcGIS Explorer, and Adobe Photoshop.

As we discussed in Chapter 2, work stored in the Places panel can be exported to a KML file by doing the following:

1. Right-click on a folder or place in the Place panel, and click *Save Place As…*
2. When the dialog box appears, navigate where you would like to save it

3. In the *Save as type* dropdown, select whether to save it as a KML file, or as a compressed KMZ file (which we will discuss next)
4. Click *OK*

Prior to saving your work as a KML file, it is important that you are organized. When data is exported, anything in the folder you selected will be exported. As we saw in these steps, you will also notice that there are two types of files you can save work to:

- *KML* files, which is used to store simple data, and will not include any embedded images you might have used.
- *KMZ* files, which are compressed files containing your geographic data and any external content you have overlayed. This can include maps, photographs or any other content you created and saved. In using this, all of your work will be stored in a single file.

A KMZ file uses the ZIP format to archive and compress multiple files used in your work. Because it is compressed, the overall size will be smaller than the KML data, images and other files that your placemarks might have referenced. When a KMZ file is opened in Google Earth, the application will automatically inflate the contents and import them.

By saving the information from Google Earth as a KML or KMZ file, you can then share them with others, allowing them to view what you want to present. For example, if you go to the Los Angeles Times Mapping L.A. Boundaries API (http://boundaries.latimes.com/sets/), you will find a list of different KML files that can be downloaded and viewed in Google Earth. These include information about the Los Angeles Fire Department, Los Angeles Police Department, LA County Sheriff, census information, and more. By downloading any of these files and opening them in GE, the data you are interested in will appear as a new layer in the 3D viewer.

## Opening KML Files

Once you have downloaded or saved a KML/KMZ file, it can easily be imported into Google Earth. Adding KML files to Google Earth is a simple matter and can be accomplished in a couple of ways. The first and perhaps easiest is to drag and drop the file into the 3D viewer or into the *Places* panel. The other is by using the menu system as follows:

1. Click on the *File* menu, and then click the *Open* menu item
2. Navigate to the location where you have saved the file, and select it
3. Click the *Open* button

When a KML file is opened (through the menu, or dragging and dropping it into the viewer), it is initially added to the *Temporary Places* folder. Anything in the folder will only remain there for the current session of using Google Earth.

Therefore, if you want it to be there after you have shutdown and restarted Google Earth, you will need to drag-and drop it into the *My Places* root or another folder.

## Viewing Code in KML Files

To view the KML code associated with a location, you can right-click on a placemark and then click *Copy* on the menu that appears. This will copy the associated KML to the clipboard. By opening a text editor, such as Notepad in Windows, you can then paste the code into the editor and see the code. A similar way to view the KML for a location is by using the Search panel in Google Earth. To illustrate this, follow these steps:

1. In the *Search* panel, enter your home address (or another location you are familiar with) into the search box, and then click the Search button
2. When your address appears in the listing below the search field, right-click on it and then click *Copy as KML*
3. Open Notepad, click on the *Edit* menu, and then click the *Paste* menu item

Once the code is pasted into a text editor, you should see something similar to that shown in Figure 4.3.

In looking at the code in the KML you pasted into a text editor, you should recognize information related to the address you entered. You will notice the data between <address> and </address> contains the address you entered, and this same information or another name appears between <name></name>. If you scroll down, you will see coordinates for this information appearing between the opening and closing <coordinates> tags, as well as other information related to the location like through fares, postal code, and so on. While some of this should not be modified in the editor (like the coordinates of that location), you could modify other data in the file. For example, change the text

Untitled - Notepad

File Edit Format View Help

```
<?xml version="1.0" encoding="UTF-8"?>
<kml xmlns="http://www.opengis.net/kml/2.2" xmlns:gx="http://www.google.com/kml/ext/2.2" xmlns:kml="http://www.opengis.net/kml/2.2" xmlns:ato
<Placemark id="21.1.3">
	<name>Trans Allegheny Lunatic Asylum</name>
	<address>71 Asylum Drive, Weston, WV, United States</address>
	<phoneNumber>(304) 269-5070</phoneNumber>
	<snippet>71 Asylum Drive, Weston, WV, United States</snippet>
	<description><![CDATA[<!DOCTYPE html><html><head></head><body><script type="text/javascript">window.location.href="http://maps.google
	<styleUrl>#listing_A</styleUrl>
	<ExtendedData>
		<Data name="placepageUri">
			<value>http://maps.google.com/maps/place?latlng=39038549,-80471694,6282576158499676030&amp;hl=en&amp;gl=us&amp;ui=ear
		</Data>
	</ExtendedData>
	<MultiGeometry>
		<Point>
			<coordinates>-80.471694,39.038549,0</coordinates>
		</Point>
		<LinearRing>
			<coordinates>
				-80.47421199999999,39.036031,0 -80.47421199999999,39.04106700000001,0 -80.469176,39.04106700000001,0 -80.4691
			</coordinates>
		</LinearRing>
	</MultiGeometry>
	<AddressDetails Accuracy="8" xmlns="urn:oasis:names:tc:ciq:xsdschema:xAL:2.0"><Country><CountryNameCode>US</CountryNameCode><Administ
</Placemark>
</kml>
```

**FIGURE 4.3** Example of KML.

between <name></name> to "My house", save the file as myhouse.kml, and then open it in Google Earth. In doing so, a placemark appearing in the 3D viewer for this location will be displayed as "My House."

Before we delve deeper into KML, and see what else we can do by editing such placemarks, let us look again at the KML you have copied into a text editor. You will notice that the first line indicates that it is actually an XML document. The reason is that KML is an XML notation that is designed to express geographic annotations and overlay visualizations. Because it is XML-based, prior to continuing our discussion on KML, we will take a look at XML and how it is structured as a markup language.

## XML

XML stands for eXtensible Markup Language. It is an official recommendation of the World Wide Web Consortium, and similarities and dissimilarities to another widely used markup language that we discussed earlier: HTML. Like HTML, XML is a meta-language that allows for the creation and formation of document markups. Unlike HTML, XML is also used in many applications and utilities and even shows up in mobile devices such as an iPhone or a GPS device like a Garmin Nuvi. While HTML is used for the presentation of data, allowing you to format content in specific ways, XML is different in that it is designed to describe data that is used by an application, and not display it.

So how is understanding XML going to help you with KML and by extension using Google Earth? Because KML is XML-based, and in fact could even be called an extension of XML, if we learn the building blocks of XML we can carry this over into KML. And knowing KML will help you tweak what Google Earth displays.

### An Introduction to XML Code

XML documents are used to store information. Tags are used to surround the information, and describe what the data is. An application looking at the XML code looks at these tags, and is programmed to know what these tags mean and how the related data should be used. To illustrate how XML works, let us look at a very simple example of code:

```
<?xml version="1.0" encoding="UTF-8"?>
<contactinfo>
<name>John Smith</name>
<company>I.M. Foremans</company>
<address>76 Totter's Lane</address>
<email>jsmith@tardis.gal</email>
</contactinfo>
```

In looking at the tags in this example, you can see that the first line indicates that it is an XML document. An application using the XML file might read the file, and use information in it to display a list of contacts, or to send email. The <contact-info> and </contact-info> tags are used to encapsulate related information together, similar to a single record in a database. Our mock program would open this file, and be programmed to look through the tags to find contact information within this section of code. In looking at the self-descriptive tags between <contact-info> and </contact-info>, you can see how the program could read the XML to retrieve the name, company, address and email address of a contact.

An application might also use an XML file to store configuration data needed for it to run, or to store options and preferences related to the application. In looking at a simple config file used by a Web application, we can see in the following code that there is a section read by the Web server that controls whether users visiting the site will be able to browse the directory structure of the site.

```
<?xml version="1.0" encoding="UTF-8"?>
<configuration>
    <system.webServer>
        <directoryBrowse enabled="true" />
    </system.webServer>
</configuration>
```

While the markup languages we discussed use a tag structure to wrap around data that is used, there are a number of differences in how they are used. While HTML uses predefined tags (e.g., <body>, <p>, etc.) to describe what certain information is and how it is to be presented, this is not the case for XML. XML has no predefined tags, meaning that you can create ones yourself. While the browser is used to display HTML tags, an application running on your computer or accessed from a Web server via a Web browser is used to process XML files.

## Basic Features of XML

Regardless of the XML document being used, there are a number of common features that can be seen in how the markup language is used.

*Elements.* An XML document is made up of one or more elements. As we discussed earlier, when we talked about HTML, the elements themselves are comprised of two "tags" – an opening and a closing tag. The opening tag is the name of the element between "less than" and a "greater than" sign as in the following – <latitude>. The closing tag is the same as the opening tag with one addition – the forward slash, "/" before the ">" symbol. An example of an element is <latitude>42.278</latitude>.

Elements can also be empty as in the following <picture src="goldengate.jpg/>. This is a short hand version eliminating the need for a closing tag by simply placing a "/>" at the end of the element.

*Attributes*. Elements can be modified by use of an "attribute" – this applies to empty elements as well. This gives additional information to the application parsing the XML to assist in its markup. An example of an element with an attribute is the following <ip="192.168.0.1" timeout="124.564000"/>.

*Document Type Definition (DTD)*. The DTD is an external document that governs how all the elements, attributes and other data of the XML document that it references are related and defined. It is not unlike a cascading style sheet in HTML. DTD references can be recognized as per the following example and will appear near the top of the document after the declaration (covered later) – <!DOCTYPE html PUBLIC "-//W3C//DTD XHTML 1.0 Transitional//EN" "->http://www.w3.org/TR/xhtml1/DTD/xhtml1-transitional.dtd">

*CDATA and PCDATA*. These are special terms that stand for character data and parsed character data respectively. When used, the former indicates that anything that occurs after the term is not to be marked up. CDATA can occur in the DTD or on its own as per this example – <![CDATA[<long>-83.733</long>]]> which a parser would not markup and display as <long>-83.733</long>. PCDATA is used to indicate that any character data that is not in itself an element can appear between the tags. For example given the declaration <!Element string (#PCDATA)> in the Apple DTD – >http://www.apple.com/DTDs/PropertyList-1.0.dtd – any character data like "http://www.google.com " or "1234" can appear between <string><string/> but not an element like <key>.

### Some Special XML Syntax

*Declaration*. Every XML document begins with a declaration about itself. This typically takes the form of something like the following <?xml version="1.0" encoding="UTF-8" ?>
*Comments*. XML has a facility for allowing comments to the code for better understanding. Anything between the tag construct <!...-> is ignored.
*Tree Structure*. XML has a tree structure, starting at a single root element – the "parent" of all the other elements and branching out to the lowest level of the tree. All elements in the XML document can have sub-elements. This is shown in the following example:

```
<root element>
     <child element>
       <subchild element>XML</subchild element>
   </child element>
</root element>
```

**TIPS AND TRICKS**

*Whitespace in XML*

You may have noticed the indentation in the above example. Though the XML specification says the whitespace in between tags is to be ignored (http://www.w3.org/TR/REC-xml/#sec-white-space) it makes it much more readable – and parse-able by a human forensic examiner – if there is indentation.

Lastly, let us briefly summarize the rules for well-formed XML

- All element attributes must have quotation marks
- All elements must have a closing tag
- XML tags are case sensitive
- XML elements must be properly nested
- XML Documents must have a root element
- Whitespace is preserved in XML
- XML stores a new line as a line feed

While this is not an all-inclusive tutorial on XML, it should give you some basics to now look at and interpret XML documents where you find them in your forensic endeavors. It will also help you understand what we are really interested in – KML. Interested readers are encouraged to look at the resources listed at the end of this chapter for further sources of information on XML.

## KML REVISITED

In this chapter, we have discussed how KML is a file format used to display geographical data, and used in browsers such as Google Earth and Google Maps (both computer and mobile). KML is based on the XML standard and like XML uses tags with elements and attributes. Also like XML, KML is case sensitive and adhere strictly to the KML reference which dictates optional tags and how within each element the order tags must appear (see the section at the end of this chapter for more information, inclusive to where you can learn more about KML and find a KML reference).

As we saw earlier in this chapter, if you ever want to see the KML code underlying any geographical feature in Google Earth, simply right click and select copy from the menu. After opening a text editor and pasting the contents of the clipboard, the KML code will be available for your viewing pleasure. Google Earth automatically creates the underlying KML code for any placemark, etc, authored from within the application. However, there are times though when you might want to tweak a specific feature or have more control over what is displayed and to do that you need to have a working knowledge of KML. This will also help you search for deleted KML code if needed.

## Basic KML Tags

The following subsections are part of the basic feature set of KML. We will be creating a sample KML file that you will be able to view in Google Earth so make sure to launch both Google Earth and your favorite text editor.

### *Placemarks*

A placemark is the most common feature used in Google Earth. It simply marks a position on the surface of the Earth with a yellow pushpin as its icon. There are three types of placemarks - simple, floating and extruded. Let us create a simple placemark - enter the following code into your text editor

```
<?xml version="1.0" encoding="UTF-8"?>
<kml xmlns="http://www.opengis.net/kml/2.2">
     <Placemark>
       <name>Simple Placemark on Lincoln Memorial</name>
       <description>This placemark is clamped to the
ground.</description>
       <Point>
          <coordinates> -77.0503,38.8892,0 </coordinates>
      </Point>
     </Placemark>
</kml>
```

As we discussed earlier in this chapter, our first line is the XML declaration or header. Because KML uses the XML standard, this is the first line in every KML file. Nothing appears before this declaration.

Line two is our KML namespace declaration. All KML 2.2 files have this line included. The KML namespace follows the same rules for an XML namespace, and used to provide for unique elements and attributes in an XML document.

Finally, in line three we get to the KML "Placemark" object. The Placemark object contains several elements that further define it:

- *Name*. This is the label that appears on the placemark when displayed.
- *Description*. This is the text that appears in the callout "balloon" describing the placemark
- *Point*. This position of the placemark on the surface of the Earth. Notice that the point element has child element of its own "coordinates". The coordinates child element specifies the longitude, latitude and altitude of the point (they must occur in that order - refer to the KML reference documentation - https://developers.google.com/kml/documentation/kmlreference#coordinates, for further information). The altitude value of "0" means that the placemark is clamped to the ground.

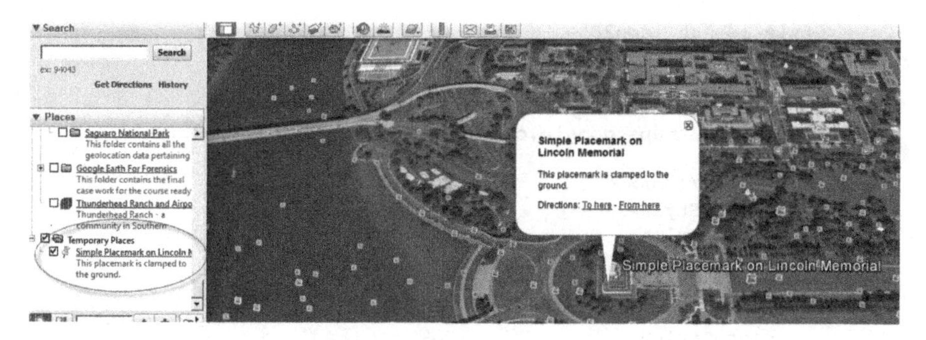


**FIGURE 4.4** Simple placemark.

If you have not already saved the sample code we provided, save the file as "simple placemark.kml". Now open Google Earth up and drag and drop the file you just created into the pane showing the earth. Your placemark should show up in Temporary Places and Google Earth should fly to your placemark. Once it centers on the placemark you can click it to bring up the balloon.

## TIPS AND TRICKS

*Editing KML With Text Editors*

Some text editors add unseen markup to a file (this occurs with RTF editors such as TexEdit on the Mac). If you have typed in the correct code for the example, have double checked for mistakes and are generating an error, try using a text editor that does not add markup to the file such as Notepad on Windows or TextMate for the Mac.

Now that we have created a simple placemark (Figure 4.4), let us work on a floating placemark. Enter the following into a new file in your text editor.

```
<?xml version="1.0" encoding="UTF-8"?>
<kml xmlns="http://www.opengis.net/kml/2.2"
xmlns:gx="http://www.google.com/kml/ext/2.2"
xmlns:kml="http://www.opengis.net/kml/2.2"
xmlns:atom="http://www.w3.org/2005/Atom">
<Document>
     <name>Floating Placemark Example</name>
     <StyleMap id="msn_icon28">
          <Pair>
               <key>normal</key>
               <styleUrl>#sn_icon28</styleUrl>
          </Pair>
          <Pair>
               <key>highlight</key>
               <styleUrl>#sh_icon28</styleUrl>
          </Pair>
     </StyleMap>
```

```
        <Style id="sh_icon28">
             <IconStyle>
                  <scale>1.4</scale>
                  <Icon>
        <href>http://maps.google.com/mapfiles/kml/pal4/icon28.png</
href>
                  </Icon>
             </IconStyle>
             <ListStyle>
             </ListStyle>
        </Style>
        <Style id="sn_icon28">
             <IconStyle>
                  <scale>1.2</scale>
                  <Icon>
        <href>http://maps.google.com/mapfiles/kml/pal4/icon28.png</
href>
                  </Icon>
             </IconStyle>
             <ListStyle>
             </ListStyle>
        </Style>
        <Placemark>
             <name>Floating Placemark - Highway</name>
             <description>The icon floats above the point at a
predefined distance.</description>
             <LookAt>
                  <longitude>174.7782340885043</longitude>
                  <latitude>-41.28922274409678</latitude>
                  <altitude>0</altitude>
                  <heading>89.99868162967627</heading>
                  <tilt>44.99932794159714</tilt>
                  <range>718</range>
                  <gx:altitudeMode>relativeToSeaFloor</gx:altitudeMode>
                  <gx:balloonVisibility>1</gx:balloonVisibility>
             </LookAt>
             <styleUrl>#msn_icon28</styleUrl>
             <Point>
                  <altitudeMode>relativeToGround</altitudeMode>
                  <coordinates>174.7784114397447,-
41.28869718693321,100</coordinates>
             </Point>
        </Placemark>
</Document>
</kml>
```

In looking over the code, you will notice that there are a few new tags to cover in this example:

- <Document> This element is a container for other elements and styles. Since we are defining shared styles for icons it is needed in the file. The shared styles are detailed below. The <name> element here refers to the name of the document.

- <StyleMap> This element maps between two different styles. In our case, we are mapping between two icon styles that will show up when the user rolls the cursor over the icon. This element takes as an attribute an "id" which names the map.
- <Pair> This is a required element to <StyleMap>. It defines a key/ value pair that maps a mode (normal or highlight) to the predefined element – <styleUrl>. It has two element requirements (which can be deduced from the preceding)
  - <key> This element identifies the key (normal and highlight)
  - <styleUrl> This element references the style. If the style is local – like in our example – the "#" referencing is used. If an external file contains the style include the full URL.
- <Style> This element describes a style that can be referenced by StyleMaps and Features. Like <StyleMap>, <Style> has an attribute "id" that names it.
- <IconStyle>, <scale>, <Icon>, <ListStyle> These elements are all children of style and server to describe the style. <Icon> provides the path to the icon being used – either locally stored or via external URL. <Scale> resizes the icon.
- <LookAt> This element defines a virtual camera that can be associated with any element that has been derived from a Feature. This element positions the "camera" in relation to the object being viewed. <LookAt> has elements that are specific to it
  - <longitude> longitude of the point that the camera is looking at.
  - <latitude> Latitude of the point that the camera is looking at.
  - <altitude> This is the distance from the surface of the Earth in meters
  - <heading> The direction(N, S, E, W) in degrees starting at zero for North.
  - <tilt> this is the angle between the direction of <LookAt> lat and long and the surface of the Earth. Values here range from 0 – directly overhead – to 90, along the horizon.
  - <range> This is the distance in meters from the point (lat, long and altitude) to the <LookAt> position.
  - <gx:altitudeMode> This tag allows for altitudes relative to the sea floor.

If you have not already saved the code you have entered into a text editor, save the file as "floating placemark.kml". Now open Google Earth up and drag and drop the file you just created into the pane showing the earth. Your placemark should show up in temporary places and Google Earth should fly to your placemark. Once it centers on the placemark you can roll over the icon to make it change per the styles you wrote into your file.

Finally, let us look at the last type of placemark – the extruded placemark. To create an extruded placemark, all that is needed is the <extrude> element placed under <Point>. This will create a visible line from the icon to the ground. Change your code for the floating placemark like the below.

```
<Document>
     <Name>Extruded Placemark Example</name>
```

and

```
<Placemark>
          <name>Extruded Placemark - Highway</name>
          <description>The icon floats above the point at a
predefined distance and a line is drawn from the icon to the
point on the Earth.</description>
```

and finally

```
          <Point>
               <extrude>1</extrude>
               <altitudeMode>relativeToGround</altitudeMode>
               <coordinates>174.7784114397447,-
41.28869718693321,100</coordinates>
          </Point>
```

Now save the file as "extruded placemark.kml". Delete your floating placemark by right clicking and selecting *Delete* and clicking *OK* when Google Earth asks you if you wish to delete. Drag an drop your new KML file to Google Earth and view the results. You should see a line leading from your icon to the ground, similar to that shown in Figure 4.5.

**FIGURE 4.5** Extruded placemark.

### *HTML in Placemarks*

While we have discussed how HTML can be used to enrich what is displayed in browsers, we will now look at how we can use it to enhance your placemarks with standard HTML markup tags. For this example, we will use the following code, which you could enter into a new text editor file.

```
<?xml version="1.0" encoding="UTF-8"?>
<kml xmlns="http://www.opengis.net/kml/2.2"
xmlns:gx="http://www.google.com/kml/ext/2.2"
xmlns:kml="http://www.opengis.net/kml/2.2"
xmlns:atom="http://www.w3.org/2005/Atom">
<Document>
      <name>Sagrada Familia</name>
      <StyleMap id="msn_purple-circle">
            <Pair>
                  <key>normal</key>
                  <styleUrl>#sn_purple-circle</styleUrl>
            </Pair>
            <Pair>
                  <key>highlight</key>
                  <styleUrl>#sh_purple-circle</styleUrl>
            </Pair>
      </StyleMap>
      <Style id="sh_purple-circle">
            <IconStyle>
                  <scale>0.933333</scale>
                  <Icon>
      <href>http://maps.google.com/mapfiles/kml/paddle/purple-
circle.png</href>
                  </Icon>
                  <hotSpot x="32" y="1" xunits="pixels" yunits="pixels"/>
            </IconStyle>
            <LabelStyle>
                  <color>ffffeaf8</color>
            </LabelStyle>
            <ListStyle>
                  <ItemIcon>
      <href>http://maps.google.com/mapfiles/kml/paddle/purple-
circle-lv.png</href>
                  </ItemIcon>
            </ListStyle>
      </Style>
      <Style id="sn_purple-circle">
            <IconStyle>
                  <scale>0.8</scale>
                  <Icon>
      <href>http://maps.google.com/mapfiles/kml/paddle/purple-
circle.png</href>
                  </Icon>
                  <hotSpot x="32" y="1" xunits="pixels" yunits="pixels"/>
            </IconStyle>
            <LabelStyle>
                  <color>ffffeaf8</color>
            </LabelStyle>
```

```
          <ListStyle>
               <ItemIcon>
                    <href>http://maps.google.com/mapfiles/kml/paddle/purple-
circle-lv.png</href>
               </ItemIcon>
          </ListStyle>
     </Style>
     <Placemark>
          <name>Sagrada Família, Barcelona, Spain</name>
          <description><![CDATA[This is the<b><i> Basílica y
Templo Expiatorio de la Sagrada Familia</i></b> or Basilica and
Expiatory Church of the Holy Family designed by <a href =
"http://en.wikipedia.org/wiki/Catalonia"> Catalan </a>architect
<a href ="http://en.wikipedia.org/wiki/Antoni_Gaudi"> Antoni
Gaudí (1852–1926)</a>. It was begun in in 1882 and construction
continues through today. For more information click on the<b><a
href ="http://en.wikipedia.org/wiki/Sagrada_Fam%C3%ADlia"> blue
link</a></b>.
          You can also add tables -
          <br><br>
          <table border="1" padding="1">
          <tr><td>1</td><td>2</td><td>3</td><td>4</td><td>5</td></tr>
          <tr><td>a</td><td>b</td><td>c</td><td>d</td><td>e</td></tr>
          </table>
          <br>
          and things like ordered lists
          <hr>
          <br>
          <ol><li>First</li><li>Second</li><li>Third</li></ol>
          <ol type="a"><li>First</li><li>Second</li><li>Third</li></ol>
          <ol type="A"><li>First</li><li>Second</li><li>Third</li></ol>
          <br>
          <hr>
          For more information see the <a
href="https://developers.google.com/kml/documentation/kml_tut">
Google Tutorial</a>]]>
          </description>
          <LookAt>
               <longitude>2.174470234286203</longitude>
               <latitude>41.40356182764409</latitude>
               <altitude>0</altitude>
               <heading>0.00106913343486977</heading>
               <tilt>0</tilt>
               <range>500.9542765372977</range>
     <gx:altitudeMode>relativeToSeaFloor</gx:altitudeMode>
          </LookAt>
          <styleUrl>#msn_purple-circle</styleUrl>
          <Point>
     <coordinates>2.174374496446947,41.403714720716,0</coordinates>
          </Point>
     </Placemark>
</Document>
</kml>
```

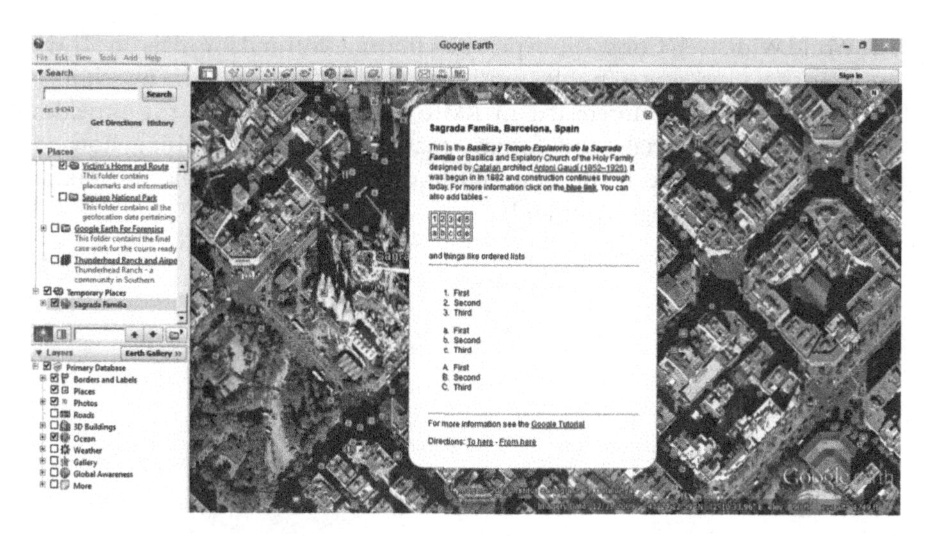


**FIGURE 4.6** HTML in description balloons.

Notice that in order to include HTML tags within the <description> element you need to use the <![CDATA[ ]]> element. CDATA stands for character data and anything inside of the element will not be treated as XML markup and hence be treated as HTML. In saving this code to a file with the KML extension, and opening it in Google Earth, you would see a placemark similar to that shown in Figure 4.6. In doing so, you would see that the HTML has enabled you to add tables, numbered and bulleted lists, hyperlinks and more.

While creating and modifying raw HTML in this fashion could require more advanced knowledge of using this markup language, keep in mind that there are HTML editors and generators that allow you to create tables, lists, format text, and so on using easy-to-use GUI interfaces. As we saw earlier in this chapter, after designing the text to appear as you want, you can then use such a tool to generate the HTML for you. Now that you know that the <![CDATA[ ]]> element is used, and how you can modify the KML data to add such content, you can modify the placemarks to show information in a more formatted and professional manner.

## LEARNING MORE ABOUT MARKUP LANGUAGES

Unfortunately, it is not possible to learn XML, KML and HTML by reading a single chapter. If you are interested in learning more about the various elements, attributes and features of these markup languages, there are a number of resources on the Internet. Using these can provide a useful reference for those already familiar with the languages, and a way to learn how to write code with these languages.

The World Wide Web Consortium provides information and tutorials on HTML on their site at www.w3schools.com/html/, and on XML at www.w3schools.com/xml/. If you are interested in KML, Google's developer site provides documentation, tutorials and forums that are useful in learning how to use KML with Google Earth. You will find information specific to KML at https://developers.google.com/kml/.

CHAPTER 5

# Digital Forensics 101

**INFORMATION IN THIS CHAPTER:**

- What is Digital Forensics
- Tools for Recovering Evidence
- Do You Really Want to Do This?
- Organizing Your Case
- Understanding What You are Looking At

## WHAT IS DIGITAL FORENSICS?

Digital Forensics is a branch of forensic science that focuses on the recovery, examination, and investigation of evidence stored on computers and other digital devices, as well as various media that may have been used to store data. Although it is commonly associated with criminal investigations, digital forensics has been used in civil cases, internal investigations, tribunals, and other inquiries or forums that require an exploration of data.

### The Process of Digital Forensics

The process of performing a digital forensic investigation can be broken down into four stages:

- *Seizure*, in which computers, mobile devices and other devices and/or media are obtained and preserved.
- *Acquisition*, in which the data is retrieved from a device
- *Analysis*, in which an image or copy of the data acquired in the previous step is examined
- *Reporting*, in which the procedures and processed that were followed in the previous steps are documented, along with the evidentiary findings

### *Seizure*

When a computer or other device is seized, it is taken into custody and secured with goal of preserving any potential evidence. As with every stage of a digital forensic investigation, you will document the scene, actions that were taken, and procedures that were followed. It is also important at this stage to establish a chain of custody that will carry on through all the other stages, documenting who and when and when a person had position of evidence.

In addition to photographing the scene where the computer or device was seized, photograph the computer or mobile device and what is displayed on the screen. Photographing the screen will preserve what applications were open, possible information, and will show what the user was last using doing on the computer or device. Under no circumstances should you use the computer/device, search for evidence, or alter its running condition. A rule of thumb is that if it is turned off, leave it off; if it is turned on, leave it on.

During the seizure, some steps may be taken to acquire digital evidence. If a computer is turned on, you would start by collecting any live data, inclusive to taking an image of the physical memory. A utility that can be used to image the RAM is F-Response (www.f-response.com). This tool could also be used to collect a logical image of the disk if you discovered the hard disk was encrypted. You would also gather any other data that is required for the investigation about the computer's live state, such as logged on users, its network connection state, running processes, and so on.

You should also take effort in documenting how the computer or device was found. Photographs and diagrams should be made of how it was setup when found, inclusive to any cords plugged into the machine. You should also label all of the cords, and document the model numbers and serial numbers of the computer/device and any other devices attached to it. Nothing should be disconnected from a computer or device until the previous steps have been completed.

When you are ready to transport the computer/device, you should package all of the components in anti-static bags, and seize any other storage media. This would include external hard disks, USB sticks, as well as CDs and DVDs that may contain data. To keep the media safe, you should avoid putting it near anything that may damage the data, such as magnets, radio transmitters, and so on. In gathering these additional items, you should also collect any manuals or documentation that may be related to the device. You never know if these will be helpful later in your investigation, or if they contain useful information (such as passwords, etc.).

There are additional considerations when a mobile device is seized. When a mobile device is connected to a cellular network, it may access new data that will overwrite evidence. Similarly, a mobile GPS unit that is turned on may continue to record track points (i.e., locations that the GPS has been) as its being transported. Because a mobile phone or tablet can be sent a command to wipe the device, you also run the risk of everything on it being erased. To preserve potential evidence on a mobile phone, GPS or other device, it is important they are stored in a Faraday bag or cage. A Faraday cage is an area protected by material that blocks signals, essentially creating the same conditions of being in a "dead zone" where you cannot get a cell phone signal from your carrier. A Faraday bag is used to store mobile devices for transport, preserving any evidence stored on them.

### *Acquisition*

The acquisition stage is where data is retrieved from a device or media, and generally occurs after the evidence has been collected, safeguarded and transported. In acquiring evidence from a device, a decision is made whether you need to perform a live or dead analysis. A live analysis is performed when a computer or device is powered on, and cannot be powered off until this information is collected. A dead analysis occurs when the machine is powered off, and transported to a lab where data can be retrieved in a controlled environment.

Acquiring data from a computer, device, or various media that may be used to store potential evidence generally requires specialized tools. This is not to say there are not times when a mobile device may require the manual acquisition of data, whereby an investigator uses the user interface of a phone or other device to view and photograph information displayed on the screen. However, in doing so, the only data that will be displayed is that which is accessible to the device's operating system and/or apps. In addition, using the interface may result in data being written to the device. To safely acquire all of the data, inclusive to that which may have been deleted, software and hardware tools are commonly used to create a bit-for-bit copy of what is stored on the device. Once a copy of the data is acquired, the investigator can then examine the copy of the data so that the original remains untouched during analysis.

There are several ways in which you may acquire a copy of what is stored on a file system, but not all of them will provide the same results. These methods include:

- Copying files, which will only copy the files that are on the system and not ones that may have been deleted. Also, metadata related to file ownership, times a file was accessed, permissions and other data may be lost in copying the file.
- Backups, which will restore a copy of the files. Depending on the backup software used, not all of the metadata related to files will be included with the backup, and it will not capture information about deleted files.

- Copying disk partitions, which will create a bit-for-bit copy of the file system including metadata related to the files and information residing in unallocated space.
- Copying the entire disk, which creates a bit-for-bit copy of the file system, including storage space before and after disk partitions.

In looking at these methods, you can see that a bit-by-bit copy of the data will yield the most possible results. While you might think this would only apply to the hard disk of a computer, many mobile devices use file systems and may be used as storage devices. In addition, devices that use SD cards can have the card removed and processed like other removable media. By using various tools discussed later in this chapter, you will be able to collect the data on these devices, making a copy that you can then analyze to identify evidence related to your case.

### *Analysis*

The analysis stage generally occurs after evidence has been collected. If live data is not being examined, then an investigation is conducted against static data that has been copied from a system. Once an image of data on the computer, device, or other media has been made, an examination of the data takes place. This may involve performing keyword searches relating to a crime, running scripts to identify certain types of data, manually reviewing information and content of files, and various other techniques.

By analyzing various types of data found on a machine, investigators will search for evidence that implicates or exonerates a suspect. The evidence may include digital photographs or downloaded images (as in the case of child pornography cases), electronic spreadsheets (in the case of financial crimes), email and other types of data. Using the content, metadata, or other information discovered, the investigator may reconstruct a series of events related to the case.

### *Reporting*

Documentation is crucial to any digital forensics case. It is important to make a record of any actions taken, devices or media examined, procedures that were followed, and other details relating to the evidence. Remember that, especially after a case goes to court, there is the possibility that anything related to the case may be questioned, and your documentation may be used to provide answers.

Throughout the process of conducting an investigation, it is vital that the integrity of the data and the device storing it is preserved, and part of this involves a documented chain of custody. Once a computer, device or media is seized, it should start the chain of custody, showing who initially took possession and who had custody of it after that point. It is also important to remember that the original devices, storage media, or other items that evidence was collected from may be requested by defense council or other parties involved in the case. In some cases, evidence files or images taken of a system may be requested. By preserving these items and ensuring there is a record of who had access to them, you can help to ensure the evidence has not been corrupted or tampered with in anyway.

It should also come as no surprise that you will need to create a report about what was found during the course of your investigation, and how it applies to the case. This could include listings and details about any files found on storage mediums (e.g., hard disks, tape, USB devices, etc.), information recovered from emails or other sources, and any other data that is being used as evidence. As we will discuss later in this chapter, many commercial tools provide features that will automatically generate reports about the files that were found. You would also write a report yourself that outlined the steps taken to acquire and analyze the data, and how the files or information found apply to the case. The reports themselves may then be submitted as evidence of an accused persons guilt or innocence.

## Where Google Earth Fits In

Google Earth (GE) can be used in multiple stages of the digital forensic process. Most often, you will find that it is used in the later parts of a case, when you need to analyze coordinates from various sources, or as a reporting tool to create presentations relating to geographic locations. In some cases, it may also be used to acquire GPS data from a device, although other tools may be more suited to collecting such data for a forensic investigation.

### *GPS Forensics*

When a person uses a GPS device, he or she will enter in locations called *waypoints* that are stored in the GPS. The waypoint may be a person's current location, or a location that he or she wants to navigate to. The GPS device will use a series of waypoints to create a *route*, showing the person how to navigate from one location to others in a specific order. Because this information can be stored on the device, it can also be retrieved and examined during an investigation.

GPS devices will also store *tracks*, which are geographic points that the unit has been. When you turn on the GPS unit, it will connect to satellites and determine its current location. As you travel, additional track points will be stored as a record of where the GPS unit has been, and stored in a *track log*. By looking at the track log, you are able to view a listing of coordinates that the portable GPS has visited and, by extension, where its owner has been.

As we saw in Chapter 3, and revisit in the next chapter, Google Earth can be used to acquire data from a Garmin or Magellan GPS unit. In performing the import, you will see the number of waypoints, tracks and routes that are imported from a GPS device, which can then be reviewed in the 3D viewer.

However, importing GPS data in this way copies the data directly off of the device into Google Earth. It does not retrieve any data that may have been deleted, or is hidden on the device.

This can be a major issue if a particular location of interested a suspect visited existed in the deleted data, and no longer appeared in the tracks you copied using Google Earth's import feature. For this reason, it is often best to use forensic tools to collect all of the data, not just what is visible to the device's interface, inclusive to any deleted or hidden data that may reside on the device.

Also, in acquiring the data from a GPS device for use with Google Earth, you want to ensure nothing is written to the GPS device. As the device will store files, your operating system or applications might write data without your knowledge or intention. If data from the original source of evidence has been modified, it could be challenged in court, and become inadmissible as evidence. To prevent this from happening, you should ensure that your forensic machine uses write protection and/or uses tools that are designed to gather evidence in a forensically sound manner, as we discuss in the next section.

## TOOLS FOR RECOVERING EVIDENCE

As we have mentioned, it is important to recognize that GE is not a tool designed for digital forensic data collection. It will do a logical download of geolocation data, so anything that is been deleted from the device (i.e., waypoints, coordinates, etc.) will not be included when you use GE to import data from the device. To acquire data in a forensically sound manner, and get all the evidence that is available (regardless of whether it is deleted or hidden), more advanced tools should be considered.

In this section, we will discuss various tools that can be used to collect data from devices. There are software and hardware solutions that prevent your operating system or software like Google Earth from writing to the device or storage media, and ones that will create an exact duplicate so that you can work from an image of the data.

### TIPS AND TRICKS

*Working with Images and Other Copies of Data*

By creating an image of what is stored on a computer or other devices, you are examining a copy of the data and not the original source. Forensic software that allows you to create an image in this way means that you can examine a computer or device without having to go through its operating system or user interface. In doing so, you are bypassing any passwords required to logon to a machine. Similarly, for mobile forensics, such tools can extract data while bypassing pattern locks, PINs or passwords.

## Write Protection

Prior to acquiring data from a GPS unit with Google Earth, you should ensure that your forensic machine has USB write protection enabled. Because a GPS unit also can function as a mass storage device, it is essential to make sure that no data on the device is changed. Rather than simply plugging the GPS device into a USB port, you want to ensure that software write protection or a hardware write blocker is used to prevent any accidental modification of data.

Write blockers allow read commands to pass from a computer to a storage device, but block any write commands. In doing so, you can safely access the drive to view its contents and\or collect data. With a hardware blocker, the disk or device you are collecting evidence from plugs into a device that becomes a midway point between the forensic workstation and the storage you are acquiring data from. The ability to block writes may also be included in other forensic hardware tools that are used to image or duplicate the data on the suspect device.

There are also a number of software solutions that can be used to prevent your computer from writing to a storage device that you are collecting data from, such as a GPS device that is connected via a USB port. On a machine running Windows, you can use write protection software like:

- DSI USB Write Blocker (http://document-solutions.biz/downloads/?did=9)
- M2CFG USB Write Block (www.m2cfg.com/usb_writeblock.htm)
- NetWrix USB Blocker (www.netwrix.com/usb_blocker_freeware.html)
- Thumbscrew (www.irongeek.com/i.php?page=security/thumbscrew-software-usb-write-blocker)

There are also a number of tools for Mac computers that provide write protection, allowing you to safely acquire data, such as:

- Softblock (www.blackbagtech.com/software-products/softblock-1/softblock.html)
- Disk Arbitrator (https://github.com/aburgh/Disk-Arbitrator/downloads)

## Tools Used to Acquire Evidence

In addition to the tools we have already mentioned, there are a number of products available for digital forensics investigations, which are commonly used by law enforcement and companies specializing in data collection. Using such suites of products, you will find that they have features and functions that will meet most of your needs throughout the process of acquiring, analyzing and reporting on digital evidence.

Guidance Software (www.guidancesoftware.com) is a company that creates a number of products used for digital forensics. The versions of *EnCase* are used to acquire evidence from hard drives, removable media (e.g., CDs, USB sticks, etc.), smartphones, tablets, GPS units and more. Using a GUI interface, the software can be used to acquire, analyze, and create reports to show what was found, where the data originated, details of files, and other pertinent facts that relate to your investigation. Once completed, you can have EnCase generate a report that can be provided to other investigators and the courts.

Cellebrite (www.cellebrite.com) is another company that is well known for its commercial digital forensic products. Using their software and hardware, you can acquire and examine data from mobile phones, GPS units, tablets, and other devices, as well as memory cards. The tools available can be used for manual acquisition, where there is a need to take screenshots or images of data, and for acquiring existing and deleted data from a device being examined.

Cellebrite also has tools specifically designed for investigations requiring the acquisition of data from GPS devices. Using these tools, you can extract data from portable GPS units like Tom Tom, Garmin and Mia, inclusive to any GPS fixes that may have been previously deleted. Once you have acquired the files using tools like Cellebrite and EnCase, you can then import them into Google Earth for further analysis.

## File Converters

While you can import GPS data into Google Earth, you are limited to files for Garmin and Magellan units. If files have been retrieved from other types of GPS devices, then you will need to convert them prior to importing them into GE. Once converted to a Garmin or Magellan format or a KML file, you can then import the data into GE. Some of the file converters available include:

- GPSBabel (www.gpsbabel.org) is freeware application that runs on your computer, which converts waypoints, tracks and routes to different formats.
- GPS Visualizer (www.gpsvisualizer.com/gpsbabel/), which is a site that provides an online version of GPSBabel, allowing you to upload and convert the file on their site.
- TraceGPS (www.tracegps.com/en/convert.htm) is another site that allows you to upload and convert files from one format to another
- GPS Data Team (http://tomtom.gps-data-team.com/poi/ov2-to-kml.php), which is a site that can convert OV2 files used by Tom Tom GPS devices to a format used by Garmin devices.

## DO YOU REALLY WANT TO DO THIS?

Just because you need the evidence does not mean that you should be the one to acquire it. Law enforcement may have a fulltime digital or computer forensic examiner, while a corporation or other organization may have someone on staff (such as in the I.T. department) who is trained in the collection of data using forensic methods and resources. Rather than doing the work yourself, you could have such a person collect the data for you, so you can work from a copy or image.

If you are not part of a formal investigation, you should ask *why* you are doing the work and where it might lead. Anyone using Google Earth has the ability to import and examine GPS data from a portable device, and retrieving and reviewing this information might be used for personal or non-investigative reasons. However, depending on what you find, that data may eventually become evidence in a court case, and how it was collected might be held to a higher standard. For example:

- A manager could import GPS data into Google Earth to review where an employee traveled during work hours. Is he or she traveling to meetings locations, customer offices and other work-related places, or visiting a bar or the beach? Looking at the GPS data would reveal where that employee goes, and if it was found the person was not doing their job, it could result in termination of employment. However, if the former employee challenged being fired and sued, then the data and methods of acquiring the GPS data could be questioned in civil litigation.
- If a friend was concerned that his/her spouse or significant other was cheating, you could examine where a portable GPS unit was taken in Google Earth. In doing so, you might confirm your friend's suspicions, but what if your findings became the basis for a divorce? What was a simple perusal of a person's goings on has now become evidence in a divorce case.

As you can see from these scenarios, a simple looksee can quickly change. When you acquire and examine any data, you should always assume that it could eventually become part of a criminal or civil case. Because of this, you should always try to follow best practices of data collection, documentation, and follow any procedures or policies created by your organization. By treating the acquisition of any data as a formal investigation, you will maintain good habits in the collection and analysis of evidence, and be prepared if you have to testify about it later.

## ORGANIZING YOUR CASE

It is a good idea to make sure that when working on a geo-forensic case in Google Earth, you make sure you keep your work organized so that it is easy to retrieve and share, that you can recover from mistakes and most importantly

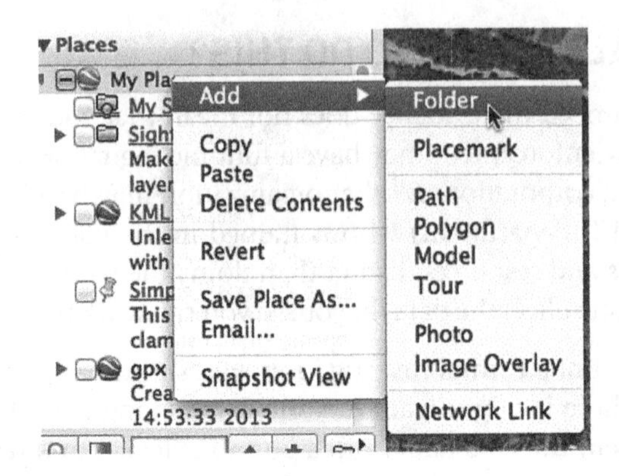


**FIGURE 5.1** Adding a folder.

you can maintain consistent work flow. A recommended way to do this is to create case folders in Google Earth. It is suggested that an investigator create two types of folders when working a case in Google Earth:

- A case folder in the "My Places" top level directory for eventual case dissemination
- A "temporary" folder in the "Temporary Places" top level directory for experimenting and developing your work.

Creating a folder in Google Earth is done by right-clicking on one of the top level directories, and when the context menu shown in Figure 5.1 appears, select *Add* and then click *Folder*.

Once you have created a folder, you are greeted with a dialog window to edit the settings of the folder. These settings are as follows.

- *Name*. Here is where you set the name of the folder. It is recommended that you use a consistent nomenclature for your particular organization. For instance <case name> - <case number>
- *Description*. You can give the folder a description of what is contains and a preview of this will appear below the folder. The description can also include links, photos and other HTML tags. This is covered in the previous chapters, and as well as Chapter 6.
- *Style, color*. This option becomes available once there are icons within the folder you are creating or any subfolders of the created folder. The option is used to create a universal color and label style in this folder and all its children.
- *View*. This option is used for creating one viewing angle for each of the placemarks contained in the folder. Once a view is set for a folder, double clicking on it will reset the view to match what was set. Setting the view will be covered in a section in Chapter 6.

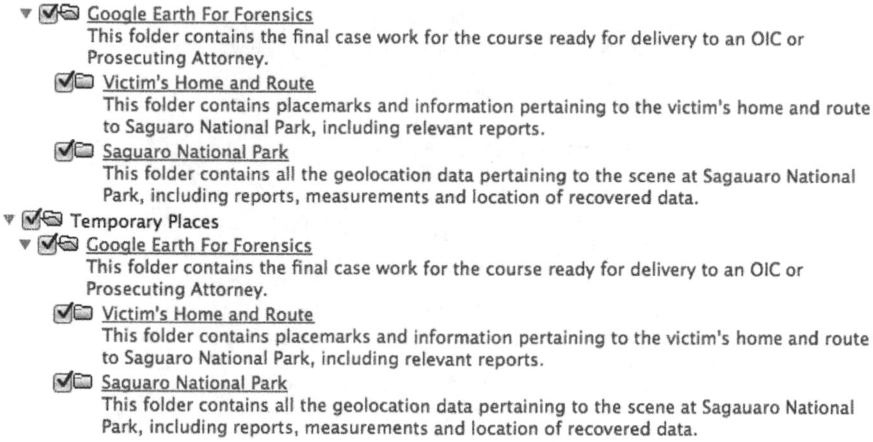

Google Earth For Forensics
This folder contains the final case work for the course ready for delivery to an OIC or Prosecuting Attorney.
Victim's Home and Route
This folder contains placemarks and information pertaining to the victim's home and route to Saguaro National Park, including relevant reports.
Saguaro National Park
This folder contains all the geolocation data pertaining to the scene at Sagauaro National Park, including reports, measurements and location of recovered data.
Temporary Places
Google Earth For Forensics
This folder contains the final case work for the course ready for delivery to an OIC or Prosecuting Attorney.
Victim's Home and Route
This folder contains placemarks and information pertaining to the victim's home and route to Saguaro National Park, including relevant reports.
Saguaro National Park
This folder contains all the geolocation data pertaining to the scene at Sagauaro National Park, including reports, measurements and location of recovered data.

**FIGURE 5.2** Folder structure template.

In Chapter 6, we will work with a scenario to use the knowledge you have acquired throughout this book. For the purposes of our scenario for this course and to get you familiar with organizing your work, create the following structure by adding folders in *My Places* and *Temporary Places*. In using this template structure, it is encouraged that you change the template and narrative contained in the description to suit the needs of your agency (Figure 5.2).

## Custom Icons

As we mentioned in Chapter 2, when creating placemarks, the *Style, Color* tab of the *Properties* dialog can be used to select a unique icon for each placemark. Using different icons makes your placemarks stand out from one another in the 3D viewer, and can provide an effective graphic representation of why a location is important and/or what was found there (e.g., a crime scene, remains, evidence, etc.).

As we will discuss in Chapter 6, you can select an icon from a library of icons that is included with Google Earth, or add a custom icon. Because you may find the ones included with GE limited, it may be useful to look at online resources, and take the time to choose ones that suit your purpose. A good site for custom icons is the Map Icons Collection (http://mapicons.nicolasmollet.com), which has hundreds of free icons that can be downloaded and used in your project. Other useful sites include:

- The Google Developers site (http://code.google.com/p/google-maps-icons/downloads/list)
- Mapito (http://www.mapito.net/map-marker-icons.html)
- Benjamin Keen (http://www.benjaminkeen.com/?p = 105)

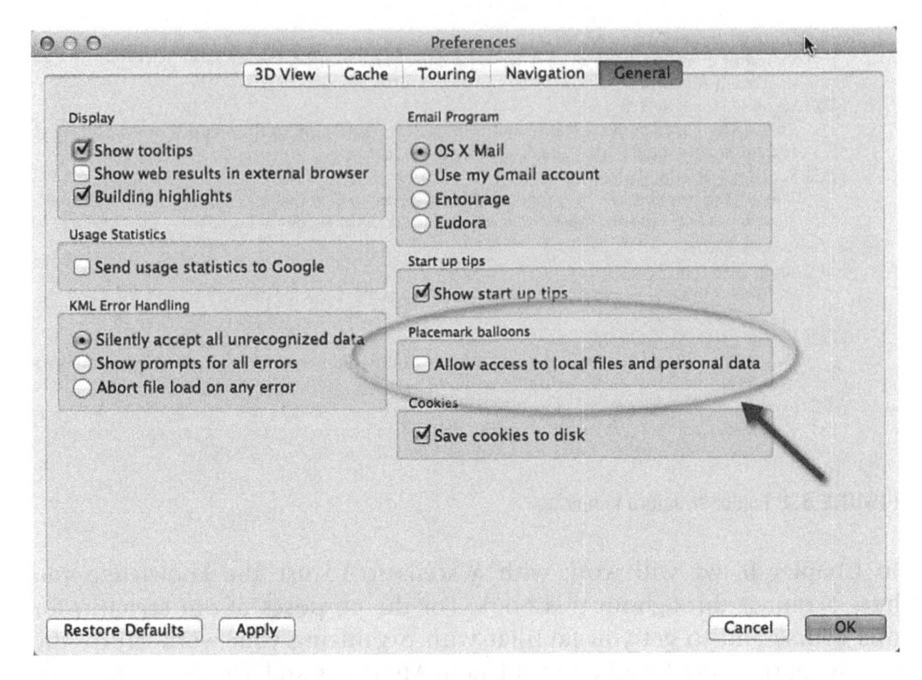


**FIGURE 5.3** Enable placemark balloon local access.

## Enabling Access to Local Files

Google Earth is set up natively to access the Internet to pull down content like map data or external files and pictures. But in digital forensics allowing Internet access by a program containing case data is generally considered to be a poor idea. It is of use, however, to use the capability of Google Earth to link to other files such as report PDFs or scene photographs. Below is the procedure for allowing Google Earth to link to files local to the examiner's machine (Figure 5.3).

1. From the *Tools* menu, and click the *Options* menu item (on a Mac click *Preferences*)
2. Click the *General* tab, and (as shown in the following figure) locate the *Placemark Balloons* section.
3. Click the *Allow access to local files and personal data* checkbox so it appears checked.
4. Accept the warning saying that access to local files might be risky, and click *OK*

# UNDERSTANDING WHAT YOU ARE LOOKING AT

When navigating through areas in Google Earth, it is important to realize that much of what is shown is not current. Some images may be recent, but others may be weeks, months or even years old. According to Google, most of the imagery you see is approximately 1–3 years old. As such, buildings that have been torn down may appear in GE, while those recently built are not visible. Similarly, the Street View does not contain real-time footage, so a familiar area may appear outdated as you take a virtual walk down the street. In using this tool, it is important to remember that what is displayed may not be an accurate representation of what is there now.

## Why is He Blurry?

In Chapter 1, we mentioned that if you notice blurred imagery in GE, it may be due to slow or poor connections to the Internet. That being said, you can expect to see some blurred areas when viewing an area in Street View. To protect a person's privacy, Google uses an algorithm that will automatically blur a person's face and the license plates of vehicles so they cannot be identified.

## Blocked Content

Generally, when you use Street View, you will not be able to access areas beyond the street. In other words, you will not be able to explore a mall's parking lot, private roads, empty fields, and so on. The reason for this is that Google uses a car with a panoramic camera on top of it to take photos as it drives down the street. It does not go off road to take photos, so you are limited to what is visible from the roadway. An exception to this is when a point of interest like Universal or Disney theme parks permit Google to enter and take digital photos of what is inside. Doing so allows you to take a virtual journey through that location.

Another time when you will notice missing content is when Google removes something that is considered inappropriate. An example of this is when you try and visit 105 Temperance Street in Manchester England, where you will find that you are prevented from navigating down a section of that roadway. The reason is that when the Google car drove by, the 15 lens panoramic camera captured multiple angles of a man and woman engaged in a sex act. The area was known for prostitution, and once it was discovered a salacious transaction had been photographed, Google blurred and later deleted the images.

## Misinterpreted Content

While Google has captured unsavory and illegal acts on camera, and even used aerial imagery showing a crime scene, there are also times where people have mistakenly interpreted what is shown. An example of this occurred on Middle Road in St John's, Worcester, England when the Google car photographed a young girl lying face down in the road, with one shoe cast off in the gutter. When the images became available the next year, users of Google Maps and Google Earth were shocked to see what appeared to be a dead girl. Fortunately, things were not what they seemed. The 9-year-old was simply playing a prank on her friend, and had been unaware that Google had snapped her picture. Before you try looking for the imagery on Google Earth, you should be aware that they have already blurred and deleted images, preventing you from navigating down that road.

## Removing Content

Problems related to what appears in Google Earth and Google Maps can be reported to the company, which may result in images being blurred, replaced or removed. To report an issue, you can use Google Maps (https://maps.google.com) to navigate to a particular location. Enter an address, and zoom into a location. When you are viewing a map or satellite image and spot a problem, you can click on the *Report a problem* link to display a dialog box that allows you to notify Google about incorrect road information, addresses, places, directions, or other issues. By clicking on the *Other Problems* link, you can report issues with satellite imagery, Street View, or other problems.

For Street View, anyone can report inappropriate content, or request that a location or person is blurred. Accessing Street View in Google Maps is the same as in Google Earth. You would navigate to a location and either zoom in as far as you can until it switches to Street View, or drop the pegman icon onto a location. Once you are in Street View, you will notice a you will see a *Report a problem* link in the lower right-hand corner. Upon clicking this, a separate browser window will open, where you can report inappropriate content. Once this window opens, you will see a picture of what you were looking at in Street View, which you can adjust to focus on a particular part of the image. You can then request that a face, your home, car or license plate, or a different object is blurred. While you have reported the issues using Google Maps, the changes will also appear in Google Earth.

CHAPTER 6

# Working a Case

**INFORMATION IN THIS CHAPTER:**

- The Practical Application of Google Earth Forensics

## THE PRACTICAL APPLICATION OF GOOGLE EARTH FORENSICS

Throughout this book, we have provided a wide variety of information on features in Google Earth, and how they can be used in an investigation. In this chapter, we will be working through a scenario to both illustrate and get practical experience with creating a case with geolocation information in Google Earth. For the purposes of our scenario it is going to be assumed that all the evidence collection and acquisition comes from a variety of sources - an Android smartphone, a Garmin GPS, geolocation information in pictures, etc. You are now in the analysis phase for the data that has been gathered.

### Details of Our Case

Our scenario is as follows.

On March 5th at approximately 5:30 AM, Barbara VanWinkle of 1920 N Forty Niner Drive left her home in her Lincoln SUV to go hiking in Saguaro National Park East near her home. Ms. VanWinkle was also an avid geocacher and as the investigation later revealed was looking for a brand new cache that was along the Tanque Verde ridge trail in the park.

Ms. VanWinkle arrived at the park and exited her vehicle approximately 20 min later at 5:52 AM. She then proceeded to hike the Tanque Verde Ridge Trail, stopping once to receive a telephone call from her husband around an hour later at 6:58 AM. The call lasted approximately 5 min and she then proceeded up the trail.



Google Earth Forensics

At approximately 7:30 AM an explosion went off at 32°10'34.51"N, 110°41'3.21"W. It was heard and seen by hikers both below and above Ms. VanWinkle on the trail. In addition the plume of dust and smoke was seen from the Park Headquarters after the explosion was heard. The hikers on the trail reached the scene at about 7:45 AM and immediately placed emergency calls.

Recovered at the scene were Ms. VanWinkle's Android cell phone, her backpack and notebook and a metal canister. The metal canister was examined at the scene by the bomb team and determined not to be an explosive device. It was opened and contained a roll note that could only be described as a "manifesto." Among other ravings the "manifesto" proclaimed war on the city of Tucson and promised more bombings. The notebook, examined preliminarily at the scene, gave the first initial clue that Ms. VanWinkle was a geocaching enthusiast because she had recorded her findings in the notebook and had journaled for the day that she was looking for a cache that had been put up the day before by a geocacher with the screen name "tardis_boffin."

The smartphone was taken back to the lab for processing.

Detectives contacted Ms. VanWinkle's husband, a professor of literature at the University of Arizona and he met them at Police headquarters. During the interview John VanWinkle exhibited erratic behavior and testimony that seemed inconsistent facts being uncovered in the investigation of Barbara's cell phone and his whereabouts. The investigation began to focus on him and police obtained a search warrant for all his electronic devices and the VanWinkle home.

The search warrant lead to evidence from Mr. VanWinkle that he had planned and carried through his wife's murder. Evidence from his laptop indicated that he had created the persona of "tardis_boffin" on geocaching.com and had created the cache that caused the death of Barbara VanWinkle. There was also evidence of Internet research regarding the creation of explosives. Mr. VanWinkle initially denied speaking to Ms. VanWinkle before her death saying, but upon examination of the professor's cell phone he recanted. There were several geotagged photographs of the scene that were taken by Mr. VanWinkle on the phone as well.

Faced with this evidence, John VanWinkle confessed to the murder of his wife. He stated that he took the pictures to help plan the murder so that it would look like a terrorist attack. He also admitted to writing the "manifesto" to try to pin the blame on a phantom terrorist bomber. He stated that he typed the note at the Bear Canyon branch of the Tucson Public Library. VanWinkle also stated that he took some chemicals from the University to build his chemical bomb and even consulted Examination of the professor's Garmin GPS turned up a "breadcrumb" trail from their home to the Library during the timespan that he stated he drove to the Library and used the computer to type the manifesto. Computer records there indicated that he had logged in using his library card.

Mr. VanWinkle stated that he wanted a divorce from his wife but she was unwilling to grant him one despite his repeatedly asking for one. When faced with her unwavering determination to stay married, he decided to kill her.

### Prior to Starting

As we discussed in Chapter 5, it is important that you organize your work as you create and modify the various files that will be used in your Google Earth forensic case. If you have not already created folders to work with in *My Places* and *Temporary Places* in the *Places* panel, you should create them now. Also, prior to doing any acquisition of data using Google Earth, make sure that write protection is used to prevent any modification of the data you might import from a GPS device.

## ACQUIRING FROM A GPS UNIT

In Chapter 3, we discussed how you can import data from a GPS unit using Google Earth. Using Google Earth in this way can be regarded as a kind of "poor mans" standalone GPS acquisition software for Garmin GPS units and some limited Magellan units. It is a logical download of geolocation data, and does not include any deleted information (i.e., coordinates, waypoints, etc.) that might be on the device. As we discussed in Chapter 5, you should consider getting a forensic examiner to acquire the data using tools designed for digital forensic data collections, and ensure the proper procedures and tools are used to safeguard digital evidence.

Now that the warnings have been given, if you are using Google Earth to import data from a GPS unit, prior to acquiring the GPS data, you should ensure that your forensic machine has USB write protection enabled (as outlined in Chapter 5) so that data on the device does not change as you acquire data from it. Once you are ready to import the data, perform the following steps.

1. In Google Earth, click on the *Tools* menu, and then click the *GPS* menu item
2. When the *GPS Import* dialog box appears (shown in Figure 6.1), select the type of device you are importing from. As our scenario involves a Garmin GPS unit, we would click the *Garmin* option to import from that device
3. In the *Import* section, select what you want to import:
   - Waypoints – these are individual locations marked on the GPS unit.
   - Tracks – these are where the GPS has been.
   - Routes – these are a series of locations where the user wishes to navigate.
4. In the *Output* section, select the output to be used with Google Earth:

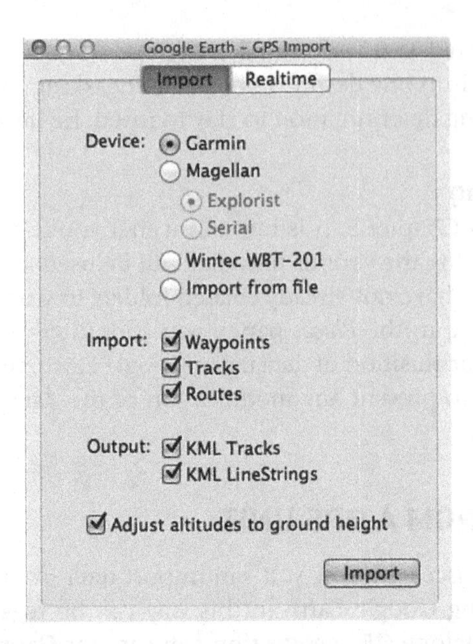


**FIGURE 6.1** Selecting GPS data to import and Google Earth output.

- KML Tracks – compatible with Google Earth and Maps versions above v5.2.
- KML LineStrings – compatible with Google Earth and Maps versions at v5.2 and below.

5. Click the *Import* button

After clicking the Import button, Google Earth will import the data from the device. You will not be given a progress bar, but there will be a dialog box asking you to wait while the program downloads the data. When it is finished, Google Earth will give you a summary of what it imported from the device (Figure 6.2).

Of interest in the previous figure is the historical timeline slider as indicated by the red arrow. This allows you to show or hide any imported data by its timestamp. For instance, in the above example moving the slider completely to the left will hide all but the data with timestamps of 8/21/2009. This is useful to narrow data to date ranges of interest.

## Importing Data From a File

If the GPS files have already been acquired from a device, then you will not be importing it from a unit plugged into a machine, but instead from a file. Importing data into Google Earth from a file follows the same steps as above with the exception of choosing the *Import from File* radio button, as shown in Figure 6.3, and then choosing the file to import.

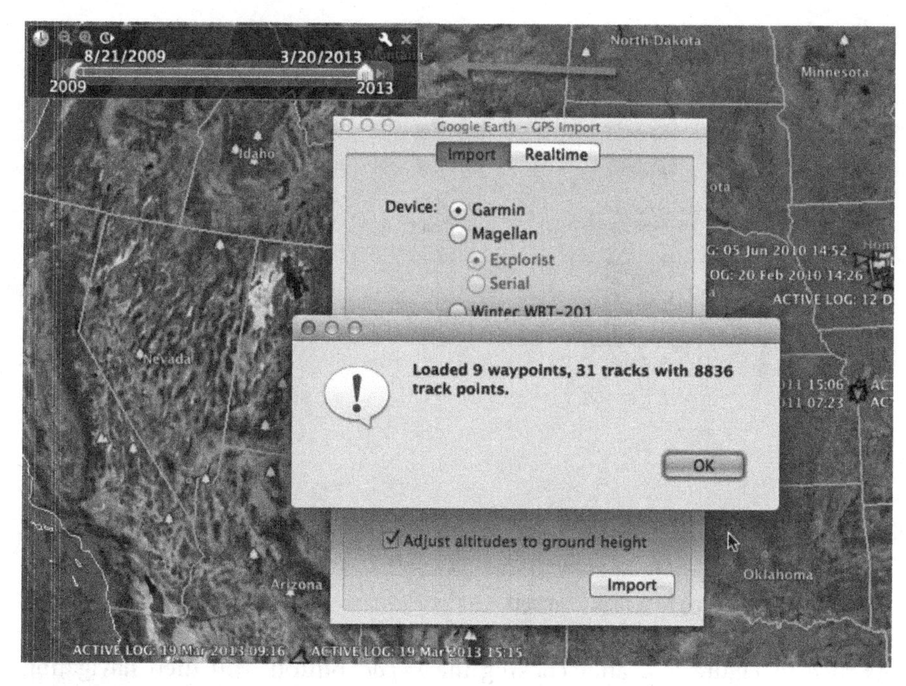


**FIGURE 6.2** GPS data from a garmin imported into Google Earth.

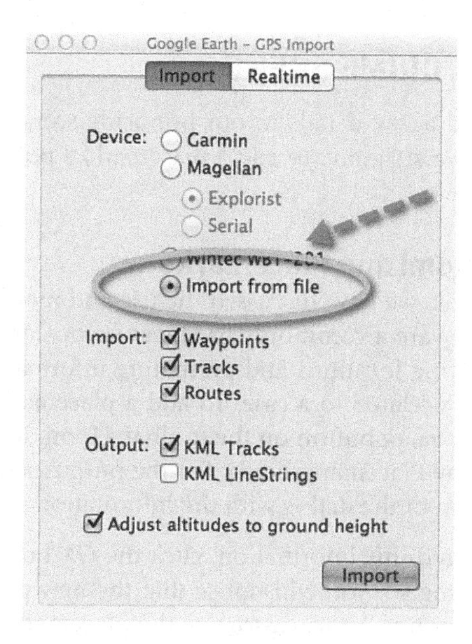


**FIGURE 6.3** Importing GPS files into Google Earth.

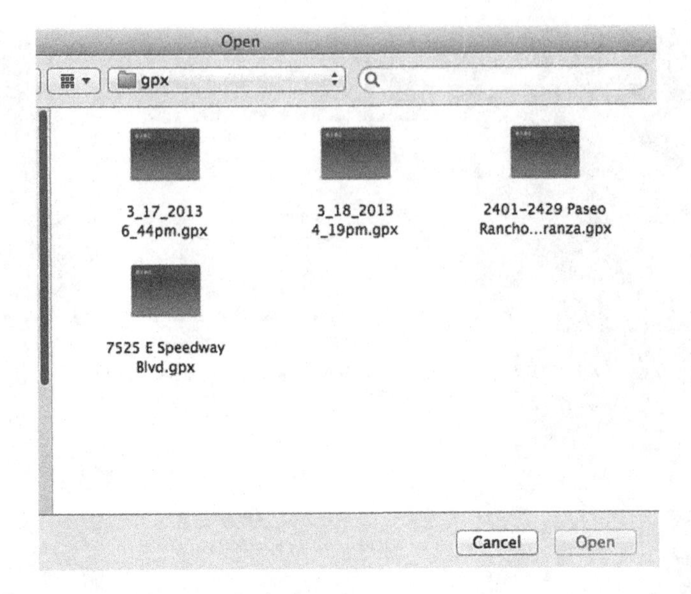


**FIGURE 6.4** Selecting GPS files to import.

As seen in Figure 6.4, after clicking the *Import* button, you then navigating to the file you want to import. Once this is done, the file will be loaded into Google Earth, and you will be presented with a prompt (which we saw in the previous section) that provides a summary of what was imported.

## ANNOTATING A CRIME SCENE

We are going to add a few details to our homicide scene through a process called annotation. We are going to add a placemark, a polygon and a path to our existing KML file.

### Adding and Customizing Placemarks

Throughout this book, we have discussed the use and modifications of placemarks in depth. They are a commonly used feature in Google Earth, and important for pinpointing locations and presenting information about a crime scene or other places related to a case. To add a placemark in Google Earth, click on the *Add Placemark* button on the toolbar. Upon doing so, you will be presented with a New Placemark dialog. For the purposes of our scenario, fill out the upper portion of the dialog with the information shown in Figure 6.5.

Once you have entered this information, click the *OK* button on the bottom of the dialog. In doing so, you will notice that the new placemark shows up under *My Places*. However, let us customize the placemark a little. Right-click on the placemark, and when the menu appears, click *Properties* (or *Get info* on

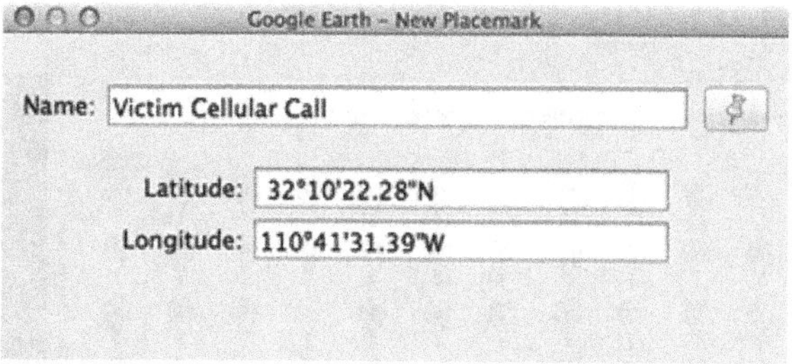

FIGURE 6.5 Add placemark dialog.

a Mac). You will now be presented with the same dialog as when you first created the placemark. Select the *Style, Color* tab. We will now proceed to change the style and color of our newly created placemark as per the numbers shown in Figure 6.6.

1. The first thing we are going to change is the icon for the placemark. As annotated in Figure 6.6 by the number 1, click the pushpin icon in the upper right hand of the window. As shown in Figure 6.7, this brings up another dialog where you can select the type of icon and adjust its size, etc. You will also notice that you can also add custom icons – from a

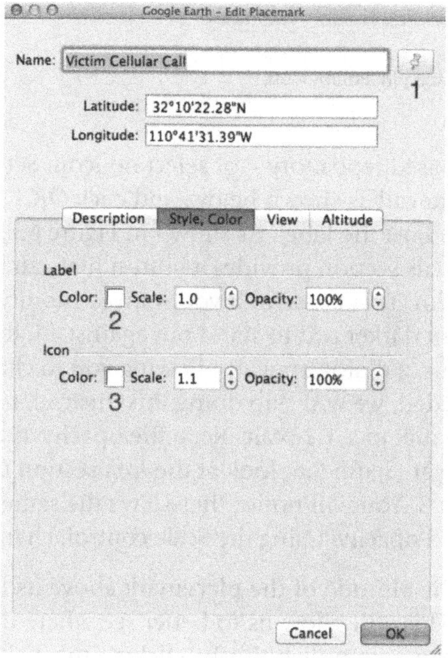

FIGURE 6.6 Style, color tab of placemark properties.

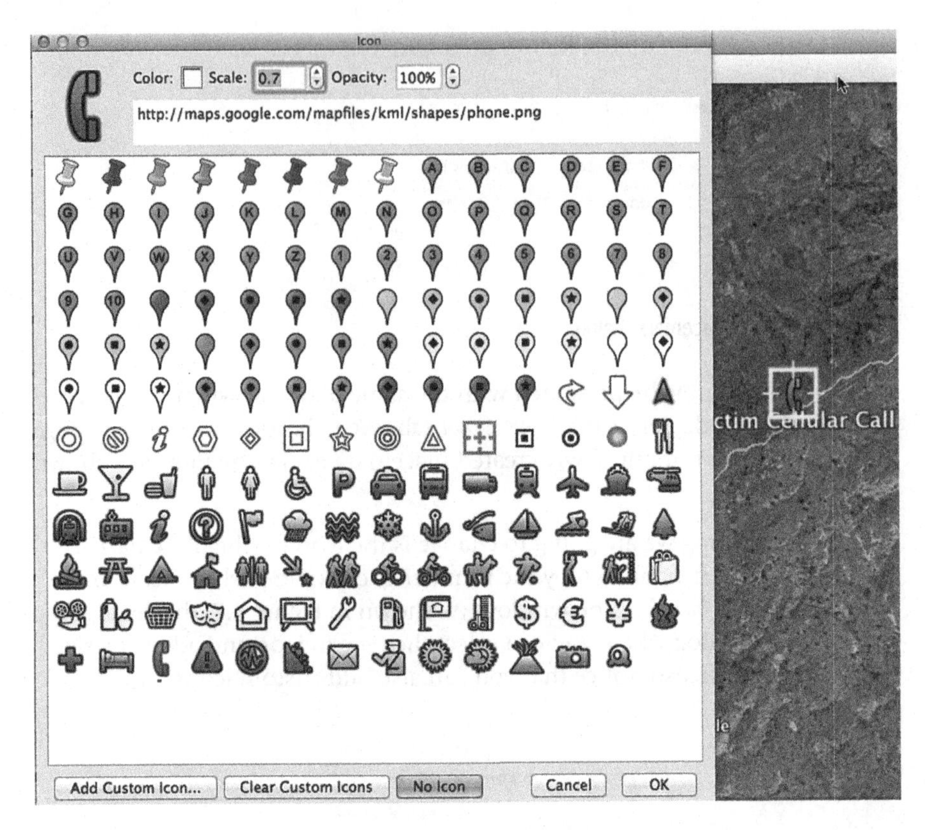


**FIGURE 6.7** Changing the placemark icon.

local or web based repository – or select no icon. Set the icon for our victim's cellular call as shown below, and click *OK*.

2. We will now adjust the label. As shown in Figure 6.6, in the section numbered 2, this section provides a button that can be clicked to change the color. This is useful if you want to modify the color (such as when you want darker text to stand out against a lighter background). If you clicked this, a dialog appears allowing you to change the color. As this is not needed, we will skip doing this. Instead, use the *Scale* control to adjust the label to a 1.1 scale. Keep the opacity as it is.
3. Again looking at Figure 6.6, look at the *Icon* section that we have labeled with a 3. You will notice these have the same options to modify color, scale and opacity. Using the Scale control, change the value to 0.7.

Now let us adjust the altitude of the placemark above its location and extend a line to its point. This will allow us to better see where this occurs along the victim's path. You can accomplish this by clicking the *Altitude* tab and putting the settings to the values shown in Figure 6.8.

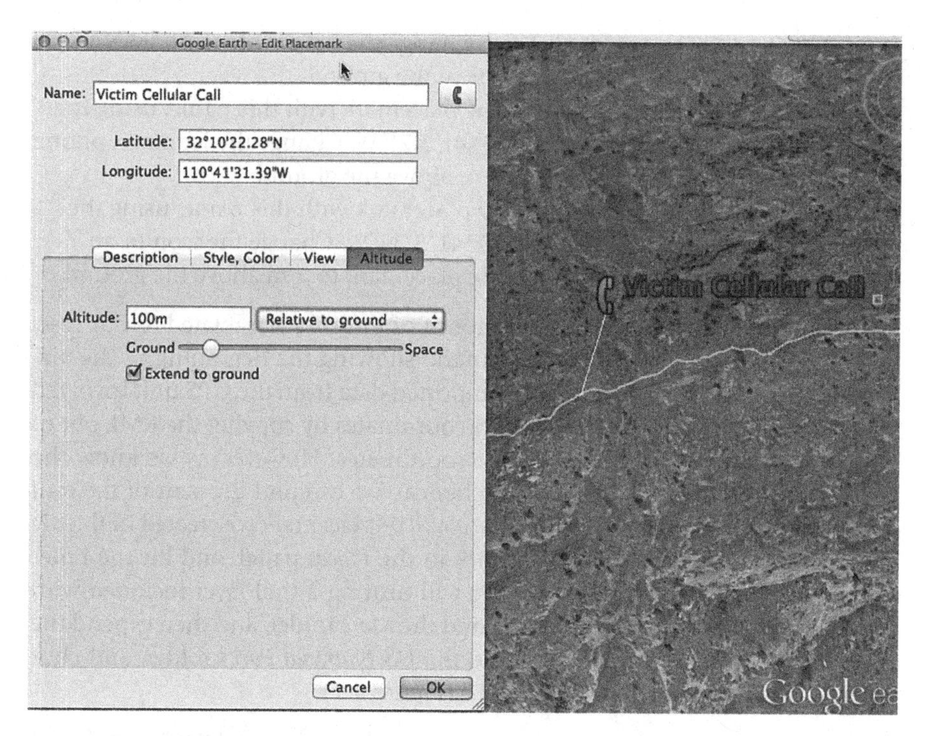


**FIGURE 6.8** Adjusting the placemark altitude.

Now that we have created this first placemark, let us look at the other placemarks related to our case. For each one, you will create the placemarks in the same way, but use some different values to make them unique to the locations we are pinpointing.

- *Victim's home.* As we saw in the scenario, she lives at 1920 N Forty Niner Drive in Tucson Arizona. You could find this exact location by using the *Search* panel, entering this address, and clicking the Search button. To save time, create a new placemark with the name "Victim's Home" and the coordinates Latitude: 32°14′43.85″N, Longitude: 110°43′59.98″W. Change the icon to the home icon and extrude the placemark to 50 m above the ground.
- *Victim's parked Lincoln SUV.* Create a new placemark with this name, using the coordinates 32° 9′56.96″N, 110°43′23.82″W. Note that the first set is latitude and the second is longitude. Change the icon to an icon of a vehicle, and extrude the icon 10 m above the ground.
- *Remains 1.* Create a new placemark with this name, using the coordinates.
- 32°10′34.78″N, 110°41′3.35″W. Change the icon to the icon with an "A" and extrude the icon 2 m above the ground.
- *Remains 2.* Create a new placemark with this name, using the coordinates.

- 32°10'34.53"N, 110°41'3.07"W. Change the icon to the icon with an "B" and extrude the icon 5 m above the ground.
- *Android Smartphone*. Create a new placemark with this name, using the coordinates 32°10'34.51"N, 110°41'3.21"W. Change the icon to a phone and extrude the placemark to 2 m above the ground.
- *Backpack/Notebook*. Create a new placemark with this name, using the coordinates 32°10'34.32"N, 110°41'3.34"W. Change the icon to an appropriate icon and extrude the placemark to 2 m above the ground.

While we have added a significant amount of information related to our case, we are going to add one more placemark, showing the beginning of the victim's hiking track. If we had actually imported data from the GPS unit involved in our scenario, we could get the exact coordinates by copying the KML out of the track and selecting the first set of coordinates. However, as we know the victim followed the trail after parking her car, we can find the start of the trail by navigating to the *Victim's parked Lincoln SUV* placemark we created in the 3D viewer. To do this, select the placemark in the *Places* panel, and hit the Enter key on your keyboard. Once there, we will turn on a trail layer included with Google Earth. In the *Layers* panel, expand the *More* folder, and then expand the *Parks/Recreation* folder. Finally, expand the *US National Parks* folder, and click on the *Trails* layer so it appears checked (Figure 6.9).

By zooming in, we can how see the area where she parked and the beginning of the trail. For this placemark, click on the *Add Placemark* button on the toolbar,

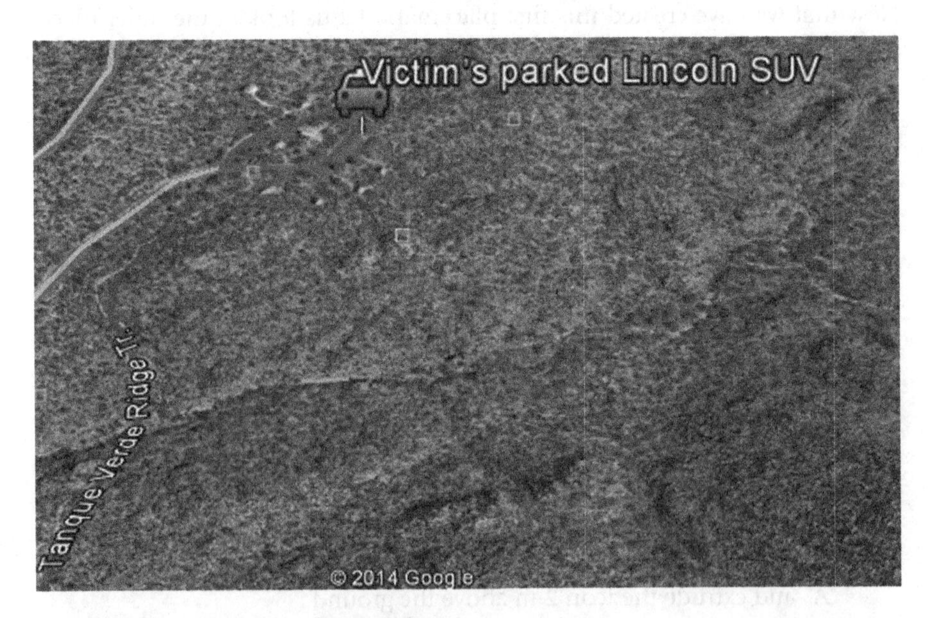


**FIGURE 6.9** Location with trail layer activated.

and notice how the new placemark has crosshairs around it. Click on these crosshairs and move the placemark to the beginning of the trail. Once done, name the placemark "Start of Victim's Hiking Route," change the default icon to the hiking icon, and change the scale to 5. Once this is done, extrude the placemark to 2 m above ground.

## Adding Descriptive Content

We will now add content to description section of the placemark. This content is what displays in the balloon when the placemark is selected. We are going to add up to three details to the description of each placemark:

- Narrative. This will occur for each placemark.
- Links to reports. We will add links to separate reports where appropriate, such as the Android Smartphone Report.
- Photographs. We will add crime scene photographs where appropriate.

In Chapter 4, we saw how we can add narrative to a placemark by right clicking on one in the *Places* panel, clicking Properties, and entering information on the Description tab. Once you have added the description, it will appear under the placemark in the *Places* panel. This can be useful in keeping track of placemarks, so you can access the right one quickly. You will also notice that Google Earth automatically creates a hyperlink in the name of the placemark, which you can click to display information.

Now that we have recapped how to do this, let us add detail to the description area of the victim's house. Fill in the description box as shown in Figure 6.10.

## Adding Local Files to the Description Tab

In Chapter 5 we discussed how you can enable access to local files on your forensic machine. With this enabled, let us add local files to the description tab so that they can be accessed in context from the placemark balloon by performing these steps:

1. Right-click on the *Victim's House* placemark in the *Places* panel, and click *Properties*.
2. Below the existing text type "Her husband was the last to see her alive and speak to her. His statement is available."
3. Place the cursor where you want the link to be added and click on the *Add link...* button as shown in Figure 6.11, and the path to your local file.
4. Click *OK* and the link will appear where you placed your cursor.

As we saw in Chapter 4, we can further modify this by editing the HTML appearing in the description box. The location of the file you want opened will appear between the quotation marks of the hyperlink tag <a href=""></a>,

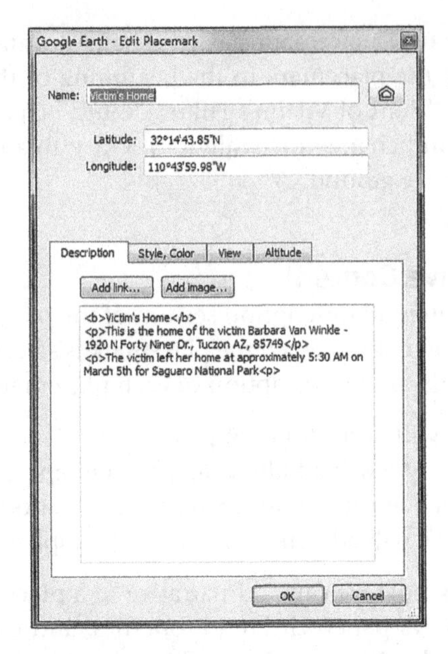


**FIGURE 6.10** Adding narrative to the description tab.

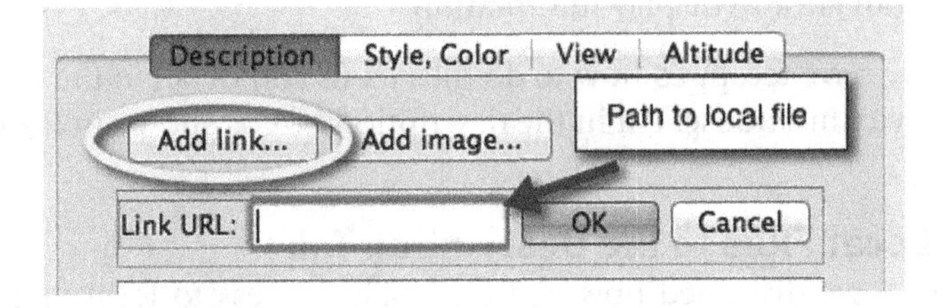


**FIGURE 6.11** Add link button.

while the text between the opening and closing tags will appear as the link users will click. When you insert a link in GE, the text for your link will also be the full path to your file (e.g., C:/casefiles/statement.doc). This should be changed to something more appropriate, so we will finish the sentence we entered earlier. Change the last sentence you entered in the description box so it appears as follows, but with *path* being the actual full path you specified for the link: "Her husband was the last to see her alive and speak to her. His statement is available <a href="file:///*path*">here.</a>. This will cause "here" to be the hyperlink as shown below (Figure 6.12).

**FIGURE 6.12** Add link button.

## TIPS AND TRICKS

*Removing Directions*

You can remove the "Directions:" line at the bottom of the balloon by a little bit of KML editing. In the KML, you will see <Style></Style> tags. Above the first <Style> tag in the code, add the following. This will add an additional style that will suppress the functionality of offering directions to and from that location.

```
<Style id="myBalloonStyle">
  <BalloonStyle>
    <text>$[description]</text>
  </BalloonStyle>
</Style>
```

After adding the style, you will need to scroll down in the KML file and replace the <styleUrl>, with the following so your placemark to reference the new style.

```
<styleUrl>#myBalloonStyle</styleUrl>
```

## Adding Photos

Adding photos and photo overlays to your Google Earth cases can enhance the presentation of the evidence. We are going to look at adding a photo overlay to the homicide scene location and then adding a photo to a location.

### *Image Overlays*

An image overlay is akin to a layer in Google Earth. It can be activated and deactivated for display over the area you are seeking to overlay. The following steps will take you through the process of adding a map overlay to the homicide scene.

1. In the *Layers* panel, expand the *More* folder, and then expand the *Parks/Recreation* folder. Finally, expand the *US National Parks* folder, and click on the *Park Boundaries* layer so it appears checked. This will help you locate the borders of Saguaro National Park East.
2. On the toolbar, click on the *Add Image Overlay* button. You can also add an image overlay by clicking on the *Add* menu and clicking Image Overlay (Ctrl-Shift-O), or by right clicking on the *Places* panel and choosing *Add | Image Overlay*.
3. When the *New Image Overlay* dialog box appears, you should also see a green outline in the 3D viewer. In the *Name* textbox of the dialog box, name the overlay "Saguaro National Park East Map"
4. In the *Link* box, type the location of the image file you want to use or click the *Browse* button to select an image stored on your machine. For our purposes in this scenario, you can type http://www.saguaronationalpark.com/images/Saguaro-East-Map.gif in this box. However, if it were a real investigation, it is most forensically sound to use a file on your machine, so content from the Web is not being used.
5. At this point, the map will probably appear oversized to what is in the 3D viewer. Use the marks Google Earth has placed on and around the image to resize and orientate it properly, so the park boundaries match what is shown on the boundaries shown in the layer you activated in step 1. As shown in Figure 6.13, this is done in the following ways:
   a. The triangle mark can be used to rotate the image
   b. The crosshairs can be used to move the image over the earth for better positioning
   c. The corner or side anchors can be used to stretch the image to fit. Holding the shift key down while using the side anchors will cause the image to scale from the center.
6. Using the Transparency slider beneath the link box, make the image overlay transparent enough to see the outlines of the park boundaries. To do this, adjust the opacity by sliding it to about halfway.
7. Add the following into the description box – "A map overlay of Saguaro National Park East."
8. Click *OK* to save the changes

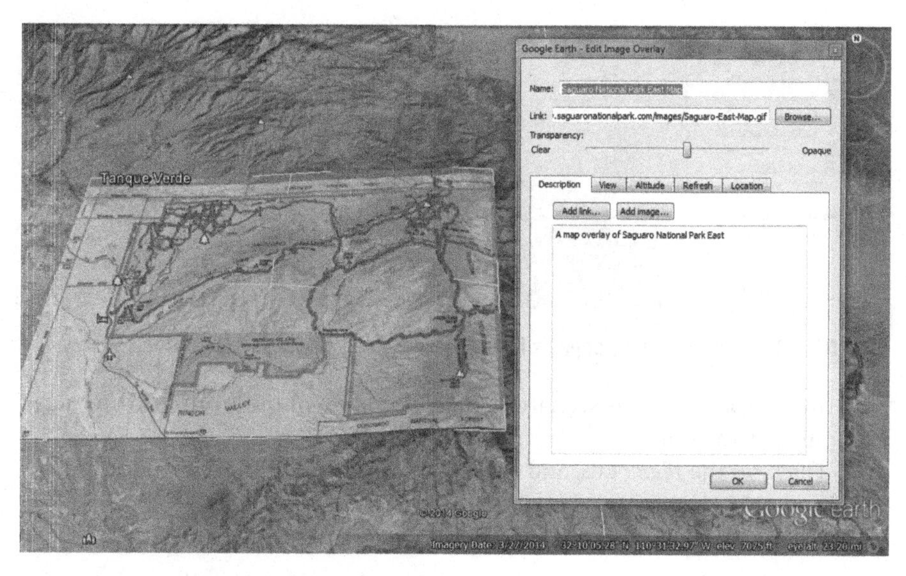


**FIGURE 6.13** Adding an image overlay.

### *Adding Photos*

Adding photos can add evidence features to your Google Earth forensic case and can enhance the presentation of your work. They can be added as an overlay in the 3D viewer, or as a picture within a placemark's description. We are going to add a photo overlay first.

> **TIPS AND TRICKS**
>
> *Photo Overlays*
>
> Photo overlays are more generally used to enhance a physical feature in Google Earth - however, here we are adding the photo to the homicide scene. Features such as being able to" fly into the photograph" are still available in our photograph, but are secondary to what we are doing - documenting what was found on the scene.

To add a photo overlay to an area, first navigate to the area where you want to add the overlay. In our case, we will navigate to the homicide scene near the placemark of that we labeled *Android Smartphone*. Once that location appears in the 3D viewer, do the following

1. Click on the *Add* menu, and then click the *Photo* menu item.
2. When the *New Photo Overlay* dialog box appears, "Victim's Phone" into the *Name* textbox.
3. In the *Link* box, enter the URL of a photo stored on a Web server, or select one from your hard disk using the *Browse* button. As mentioned, using photos from your hard disk is advisable for forensic cases, but for

**FIGURE 6.14** A photo overlay appears in the 3D viewer and places panel.

the purposes of this scenario, you could use a photo from the Internet, such as the one

4. Click on the *Photo* tab, and enter 32°10′34.51″N in the Latitude field, and 110°41′3.21″W in the Longitude field.
5. Click *OK* to save your changes.

Once the photo has been added, as shown in Figure 6.14, it will appear in the 3D viewer and in the *Places* panel. Users can now interact with it, as you can see by zooming away from the photo, and then double clicking it to be "flown" into the picture.

As we discussed in Chapter 4, you can also add photos to the description of a placemark. Adding a photo in this way is accomplished through the following steps:

1. In the *Places* panel, right-click on an existing placemark and then click *Properties*. For the purposes of this exercise, you can use the *Android Smart Phone* placemark.
2. When the dialog box appears, the *Description* tab will be open by default. Click on the *Add image…* button.
3. When the *Image URL* field appears, type the path or URL of the image you want to use. For the purposes of this exercise, you can use the same image you used in the previous exercise where you added a photo overlay called *Victim's Phone*.
4. Click *OK* to add HTML code to the description box, and then click *OK* again to save your changes.

Now that the image has been added to the description box, when you click on the placemark in the 3D viewer you will see the picture appear in the balloon. In doing so, you have provided easy access to photographs related to that placemark, such as evidence found at a particular location.

## Adding Shapes

Google Earth allows you to draw shapes on the terrain, which can help draw attention to a particular detail or set of details. By drawing a polygon onto an area, a person looking at the 3D viewer will notice the area you have designated, and can view additional information by clicking on it.

Creating a polygon in Google Earth is a relatively straight forward affair. To find the area we will be using in this exercise, type "Thunderhead Airport, Tucson, AZ" in the *Search* panel and click the search button. Once it appears in the 3D viewer, follow these steps:

1. Click the Polygon icon on the toolbar, or right-click in the *Places* panel, select *Add* and then click *Polygon*.
2. When the *New Polygon* dialog box appears, type "Thunderhead Ranch and Airport" in the *Name* field
3. In the description box on the *Description* tab, type "Thunderhead Ranch – a community in Southern Arizona south of Saguaro National Park East"
4. In the 3D viewer, you will notice that the pointer for your mouse has changed to crosshairs. Click in the viewer to being placing points for the polygon. As you do so, you will notice that the polygon will begin to take shape.
5. Lastly, click on the *Altitude* tab. Here you can change the altitude of the polygon relative to the ground. Change the altitude to 10m, and click the *Extend Sides to Ground* in order to show the polygon encircling the area of interest. Failing to check this will cause the polygon to "float in the air" and could be confusing to another user

## Measuring

Another useful feature in Google Earth is the measuring feature. All versions of Google Earth can measure lines and paths and the paid versions of the software can measure other things like circumference of polygons and 3D buildings. To make such measurements, you would start by pressing the *Ruler* button on the toolbar. You can also access the ruler by clicking on the *Tools* menu, and clicking the *Ruler* menu item. Upon doing so, a dialog box similar to Figure 6.15 will appear.

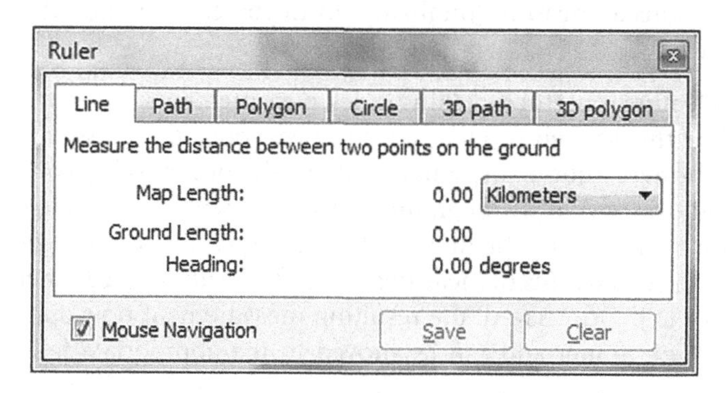


**FIGURE 6.15** Ruler dialog box.

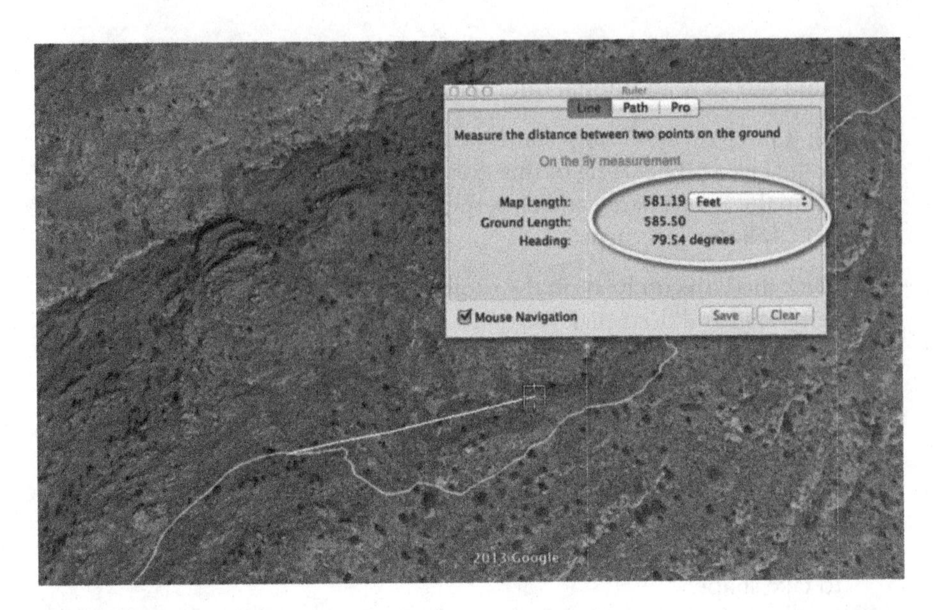


**FIGURE 6.16** Ruler dialog box.

When the dialog box appears, you will notice that the pointer for your mouse has become crosshairs. As we saw in the last section when we created a polygon, clicking on a spot in the 3D viewer will indicate where you want to start that shape from, while additional clicks will create the shape that will be measured. The particular shape you are creating and measuring is determined by the tab you have selected in the dialog box (i.e., line, path, polygon, circle, 3D path or 3D polygon). To show how this works, we will measure a line, which provides a measurement between two points:

1. With the line tab selected in the dialog box, click on the spot in the 3D viewer that you want to start the line measurement from.
2. Now move the pointer to the ending point of your measurement. As shown in Figure 6.16, Google Earth draws a line from the starting point and begins to measure "on the fly" until you click to select the ending point.

While this feature is useful for taking quick measurements between two points, there will be times when you will want to save your line measurement. To do this, click the *Save* button, and a new dialog will appear asking you to provide a name and description. Through the dialog, you will also be able to customize the line by modifying the line width and color. The *Measurements* tab also allows you to set the units of measurement to be used (e.g., feet, meters, miles, kilometers, etc.). Once saved, the resulting measurement now appears in the *Temporary Places* folder and can be moved to the appropriate folder in your case.

While a line is a measurement between two points, a path can be used for measuring multiple points. Creating a measurement for a path can be accomplished in the following way:

1. Click on the *Ruler* icon on the toolbar like above to bring up the dialog.
2. Click on the *Path* tab
3. Click in the 3D viewer. This will leave a point. Continue to click in the direction you want the path to go. Google Earth will plot the path and measure it as it grows.
4. When you are finished with your path measurement, click the "Save" button. You will be prompted to set the name and any additional customization.

As it was when we saved a line, the resulting measurement now appears in the *Temporary Places* folder and can be moved to the appropriate folder in your case.

**TIPS AND TRICKS**

*Adding Descriptions*

Adding a description to the measured path/line makes it behave like any placemark, i.e., the path will have a hyperlink that if clicked will display the balloon and double clicking on it will "fly" you to the path or line. Just like a regular placemark you can add other enhancements like pictures, HTML or unique views to enhance its display.

## VIEWS AND CAMERA ANGLES

Sometimes when creating a forensic case and presentation in Google Earth, you may want to show a unique view or perspective when someone first clicks on a placemark or folder. Knowing how to set unique camera angles and perspectives will help you to present your evidence in the manner that is most effective for your purpose. The following section explains how to set unique views and camera angles.

The first thing you will want to do is navigate to the area where you will setup a unique view. In Google Earth, select one of the placemarks for our homicide scene and hit the Enter button on your keyboard to be taken there. In doing so, Google Earth will center the view over the placemark but there is not any special angle or perspective to any of them. We will now set a unique angle for the *Backpack/Notebook* placemark:

1. Zoom into the placemark, and using the navigation tools and your mouse, adjust the tilt and rotation so it provides a different perspective on the location
2. In the *Places* panel, right-click on your placemark and click *Snapshot View*.

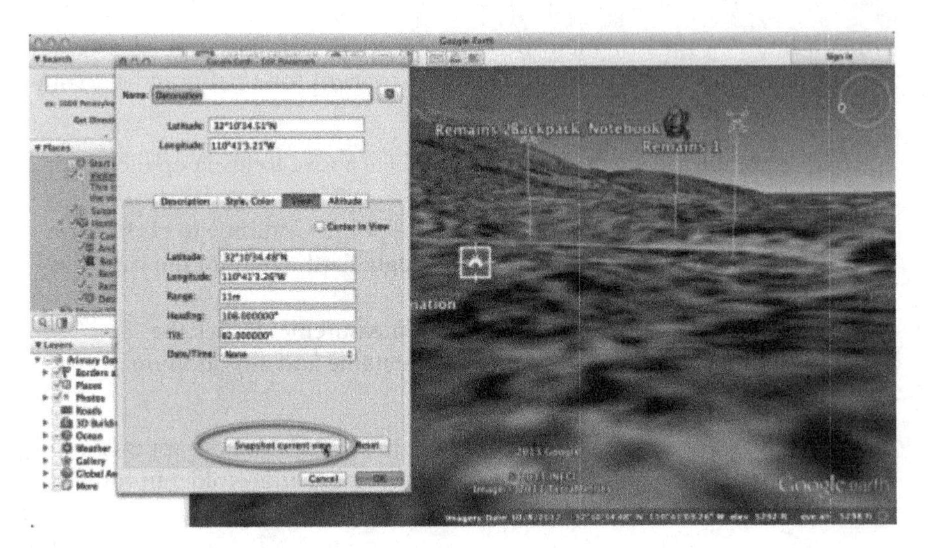


**FIGURE 6.17** View tab of placemark.

Once you have taken a snapshot of the view, when you now go to that placemark by selecting it and hitting Enter on your keyboard, you will see the placemark from the perspective you just set.

You can also set your viewpoint by using the *Properties* dialog box for a placemark. By right clicking on the placemark and clicking *Properties*, you would then click on the *View* tab to set how the placemark will be displayed in the 3D viewer. The controls for latitude, longitude, heading, range, and tilt are particularly useful when you want to add specific settings for a view. Once you have adjusted the settings to your liking, you can then click the *Snapshot current view* button to take a snapshot, and *OK* to save the view. If you decide you do not want to use the snapshot anymore, you can click the *Reset* button which will remove all of the tilt and heading settings related to the placemark, and restore it to the original view (Figure 6.17).

To keep the placemark centered in the view, you should ensure that the *Center in View* checkbox is checked. This will keep the placemark centered on the screen, so that it is not repositioned or lost from view as you drag your mouse and move around in the 3D viewer.

Setting a camera for a folder is useful to useful when you want to quickly focus onto a specific area, such as when presenting evidence or creating "tours" for display. The procedure for this is the same as for adding a view on a placemark:

1. Zoom in on the homicide scene and set an unique perspective for the entire scene.
2. Right-click and click *Snapshot View* or right-click and then click *Properties* (*Get Info* on a Mac) to bring up dialog box.

3. If you used the *Properties* dialog, then click on the *View* tab and click *Snapshot current view*. Click *OK* to save your changes.
4. Finally, zoom away from the scene and then double click the folder to test your view.

**TIPS AND TRICKS**

*Changing Settings*

Modifying the settings for a single item affects only that items display. Changing style and altitude settings for a individual folder items disables universal style sharing. However, the shared styles previously created are still applied to other items in that folder.

Because the style sharing is disabled when settings are modified for individual items, it is best to first set common styles - like line width, color and scale - on the folder level and then apply individual changes. If you again set folder style settings the shared styles are reactivated and the individual changes are lost - you will have to reapply your individual modifications.

## LEGENDS, LOGOS, AND BANNERS

Adding a legend, banner and logo to your cases can help explain details of what is being shown in the 3D viewer and can enhance the case presentation. For example a banner could be added that will indicate that the case is classified, or a legend can be added to explain details of a particular feature being highlighted.

Adding a logo, banner or legend requires a small bit of KML hacking - but it is an easy hack. Prior to modifying the KML, you should have a graphic prepared that will be used for the legend. While creating the graphic goes beyond the scope of the course, there are numerous programs (such as Adobe Illustrator) that you can use to create your custom graphics.

We will begin by adding a legend to our homicide, but before we do that we need to create the folder structure for our three new features. The following steps will walk you through what needs to be created to organize our banner, legend and logo.

1. Using the folders you created for our homicide case, create a new folder underneath the *Saguaro National Park* folder. If you did not do this, then create a new folder in the Places panel by right clicking, selecting *Add*, and then clicking *Folder*.
2. Name the folder "Banner/Legend/Logo"
3. Now that we have created our folder, we can create our KML template file that will serve for each of our three features. Open up your text editor and enter in the following:

```
<?xml version="1.0" encoding="UTF-8"?>
<kml xmlns="http://www.opengis.net/kml/2.2"
xmlns:gx="http://www.google.com/kml/ext/2.2"
xmlns:kml="http://www.opengis.net/kml/2.2"
xmlns:atom="http://www.w3.org/2005/Atom">
<Document>
    <ScreenOverlay>
    <Name>YOUR NAME OF OVERLAY</name>
    <Icon> <href>PATH TO YOUR FILE</href>
    </Icon>
    <overlayXY x="0" y="0" xunits="fraction"
yunits="fraction"/>
    <screenXY x="25" y="95" xunits="pixels" yunits="pixels"/>
    <rotationXY x="0.5" y="0.5" xunits="fraction"
yunits="fraction"/>
    <size x="0" y="0" xunits="pixels" yunits="pixels"/>
    </ScreenOverlay>
</Document>
</kml>
```

4. Save the file as *B_L_L_template.kml* in a location you will be able to find. We will now modify this template for the three files that we will create underneath our "Banner/Legend/Logo." First up is our legend overlay.
5. Now change the value between the <Name> tags to be "Legend: Homicide Scene" – without the quotes.
6. Change the value between the <href> tags to be the path to the graphic file you have created for your legend, which should reside on your local machine
7. Now we are going to set the location of the legend. To do this we need to alter the "*x*" and "*y*" values in the <screenXY> element. Remember that in a graph the *x* value is horizontal and the *y* value is vertical. We want our legend to be anchored in the lower left corner of the 3D viewer so we are going to set the x and y values to "0" to do that.
8. Save your KML file as Legend.kml and drag and drop it into Google Earth. It will show up under your "Temporary places." As seen in Figure 6.18, a legend should now appear in the 3D viewer. Make sure it is displaying in the manner you wish. If it is you can drag and drop it into your *Banner/Legend/Logo* folder and then delete the Legend.kml from your Temporary Places.

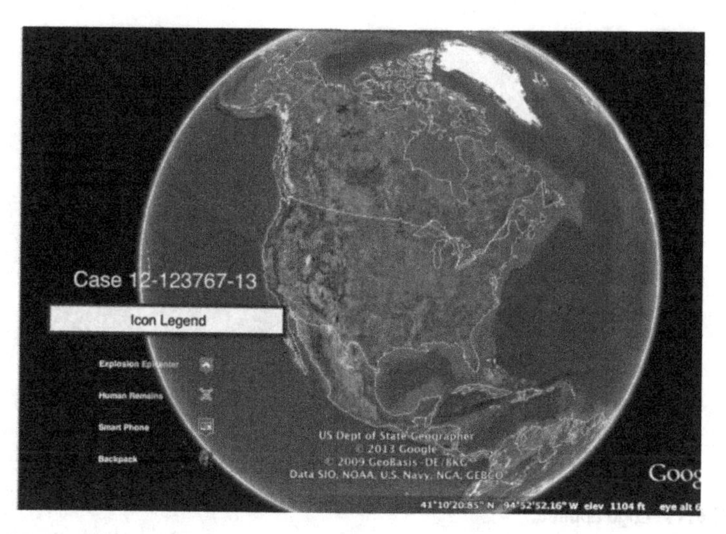


**FIGURE 6.18** Legend applied.

Now that we have the legend applied to the 3D viewer, we will create the logo file. This would be the logo of your organization, which you should be able to get as a .JPG or .PNG from the agency or company you work for. To add the logo to your project, do the following.

1. Change the value between the <Name> elements to be "Logo" – without the quotes.
2. Change the value between the <href> elements to be the path to the logo file on your local machine.
3. Now we are going to set the location of the logo. To do this we need to alter the "*x*" and "*y*" values in the <screenXY> element. Remember that in a graph the *x* value is horizontal and the *y* value is vertical. We want our legend to be anchored in the upper right corner of the 3D viewer near the navigation controls so we are going to set the x value to "900" and the y value to "500" to do that.
4. Save your KML file as Logo.kml and drag and drop it into Google Earth. It will show up under your "Temporary places". Make sure it is displaying in the manner you wish. If it is you can drag and drop the Logo layer underneath the *Banner/Legend/Logo* folder and then delete Logo.kml from your Temporary Places (Figure 6.19).

Finally we will create our "Classified" Banner for our case:

1. Change the value between the <Name> elements to be "Banner" – without the quotes.
2. Change the value between the <href> elements to be the path to the image file on your local machine that will be used for the classified banner.

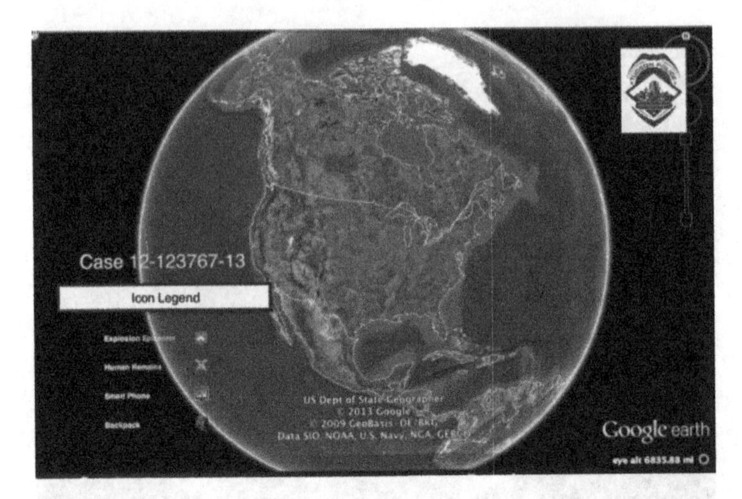


**FIGURE 6.19** Logo applied.

3. Now we are going to set the location of the banner. To do this we need to alter the "*x*" and "*y*" values in the <screenXY> element. Remember that in a graph the *x* value is horizontal and the *y* value is vertical. We want our legend to be anchored along the bottom of the 3D viewer near the right corner so we are going to set the x value to "550" and the y value to "0" to do that.
4. Save your KML file as Banner.kml and drag and drop it into Google Earth. It will show up under your *Temporary places*. Make sure it is displaying in the manner you wish. If it is you can drag and drop the Banner layer into your *Banner/Legend/Logo* folder and then delete Banner.kml from your Temporary Places (Figure 6.20).

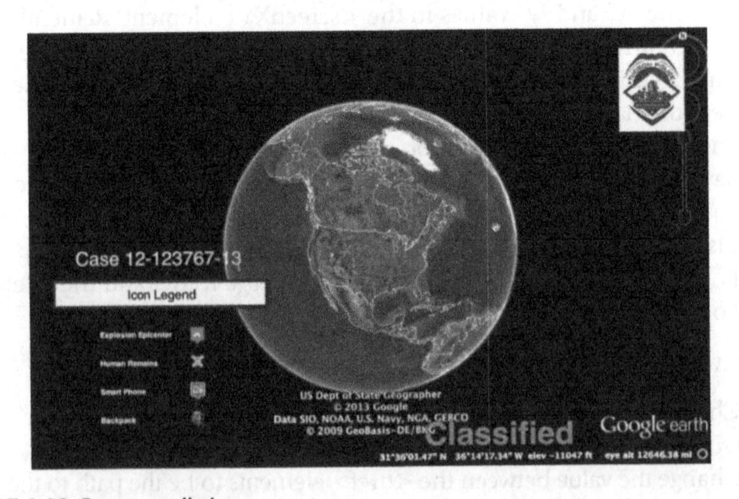


**FIGURE 6.20** Banner applied.

## CREATING A TOUR OF THE CRIME SCENE

In Chapter 2, we discussed how you can use Google Earth to make tours of locations. Creating a narrated tour of the features that you have added to your Google Earth case can greatly enhance the details of your case and the understanding of your target audience. In this section we are going to create a virtual walk-through of the homicide scene. We are also going to add voice narration to your walk-through as a further enhancement.

For now, you can turn off the Banner/Legend/Logo layer that you created in the previous section.

To create a tour of our crime scene, follow these steps:

1. Make sure you have the *Saguaro National Forest* folder selected and zoom into the location. We are going to create a walk-through that approaches the scene from the hiking trail.
2. As we saw in Chapter 2, there are multiple ways to add a walk-through to your case. Right-click within the *Saguaro National Forest* folder, select *Add*, and then click *Tour*. Alternatively, you could also click the *Add* menu, and then click the *Tour* menu item, or simply click the *Add Tour* button on the toolbar
3. When the recorder bar appears in the lower left of the 3D viewer, click the *Record* button to start recording. It will appear red indicating that you are now recording. To stop recording, simply click this button again.
4. Navigate in the direction that you want the viewer of the walk-through to proceed by using the navigation tools and suggestions covered in Chapter 2.

During your walk-through you can utilize the unique perspectives of your placemarks by double clicking on them, get rid of description balloons and turn on and off other features such as overlays. Luckily there is an easy way to create walk-throughs and add narration. Before discussing this method, we need look over the tour settings.

### Touring Settings

It is important that you take a bit of time to contemplate the touring settings when creating your walk-through of your scene. It is in the touring settings that you have control over things like the length of pause at a particular feature, whether or not balloons appear when pausing at a feature, etc. You can access these settings clicking on the *Tools* menu, clicking the *Options* menu item (which would be *Preferences* on a Mac), and then selecting the *Touring* tab that is shown in Figure 6.21.

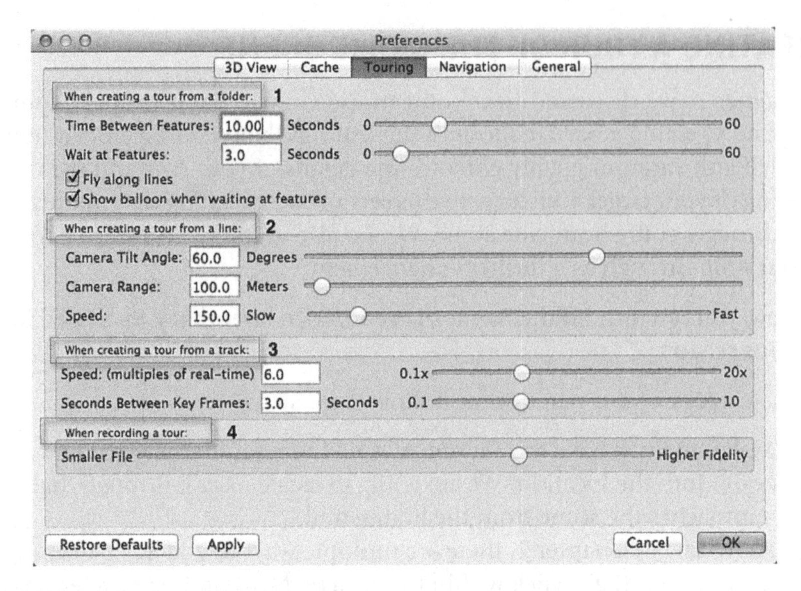


**FIGURE 6.21** Touring settings.

In looking at the Touring tab, you will see that the options are grouped together into different sections:

- The *When creating a tour from a folder* section is where you set how to create a tour or walk-through from a folder. The default setting is to have *Fly along lines* and *Show balloon when waiting at features* unchecked.
- The *When creating a tour from a line* section is used to set how the tour or walk-through will move along a line. This setting is in effect when checking "fly along lines" in the first section. Adjust the settings here to get a higher or lower viewer of a line, the speed when traversing the line and the camera tilt.
- The *When creating a tour from a track* section is used for the settings when creating a tour from a track. Tracks occur commonly from downloaded information from a GPS unit.
- The last section adjusts the size of the tour file that you create.

## TIPS AND TRICKS

*Touring Settings*

Generally speaking the time between feature default of 10 s in the first section is a bit long. Consider adjusting these two values based on how many features you have in the folder and the amount of text in the description balloons (if any).

### Creating Tours From Folders

Creating a "tour" from a folder is a handy and easy way of playing out the organized features of a scene or location. Having the features grouped together, makes it easy to record a narrative to go along with each of the features.

Remember when we worked on setting the unique views and perspectives of your placemarks and features? This work now comes in handy when we are creating a tour for disseminating and explaining our case.

Creating a tour from a folder can be as easy as the following steps.

1. Select the folder you want to create the tour
2. Locate the play tour button in the lower right hand corner of the *Places* pane and click it – your tour should play.
3. Once the tour has played you can then save it by pressing the save icon in the player

If you are not satisfied with the results of the your tour, you can go back and adjust the views and settings as needed.

### Recording Voice Over Narration

As we saw in Chapter 2, you can record a voice over to go with a tour as you are creating it. By following these steps, you can provide narration as it is playing:

1. Either create a tour or select the folder you want to create a tour on.
2. Click the *Record* button on the toolbar to bring up the recording control
3. Click the microphone icon to start a narration recording.
4. Immediately after completing step 3 press the play button on the tour toolbar to start the tour.
5. Record your narration as the tour plays.

Once you are done, playback your new recorded tour and (if satisfied) save your tour. It will now appear under the folder you created it in.

## DISTRIBUTING YOUR WORK IN GOOGLE EARTH

Once you have added all your evidence and have arranged it all in the way that you want it to look, you will want to deliver the product to the investigating officer or client or to distribute it for presentation in court proceedings. The actual distribution can be accomplished in one of several ways. Each method involves creating a KMZ archive of the case folder that can be emailed or burned to a CD/DVD/USB removable device.

As we saw in Chapter 4, creating a KMZ archive is easy:

1. Right-click on the folder or placemark that you want to share as a KMZ, and click *Save Place As…* from the menu

2. Navigate to the location you want to save it and name the file with the nomenclature appropriate to your agency.
3. You can now distribute the file.

Sharing via email is also an option, however, forensic machines *should not be* connected to the Internet. The process for sending KMZ files via email are done in the following manner:

1. Right-click on the folder or placemark that you want to share as a KMZ, and click *Email...* from the menu
2. Google Earth will then ask you what email program you want to use for emailing the file
3. Google Earth will automatically add the file as an attachment to your email

# Index


## A

Acquiring from, GPS unit, 85–88
Adding descriptive content, 93
Adding local files to the description tab, 93
Adding photos, 96–98
Adding shapes, 98
Add link button, 94
Add Placemark button, 88
  dialog box, 89
A-GPS system, 41
Android devices, 5
  smartphone, 5, 41, 92
Annotating a crime scene, 88
Application Programming Interface (API), 4
  Google Maps, 5
  JavaScript, 4
ArcGIS ArcScript, 45

## B

Backpack/Notebook, 84, 92, 101
  placemark, 101
Banners, 103
  applied, 106
  classified, 105
Blocked Content, 81
Blurred imagery, in GE, 81
Browsers, 4
Buttons, on Google Earth Toolbar, 15

## C

Camera angles, 101
Cellebrite, 76
Cell-site multilateration, 41
Changing the placemark icon, 90
Chinese BeiDou Navigation Satellite System, 36
Computer-Assisted Dispatch (CAD) systems, 43
Configuration, 28
  Cache, 31
  3D View, 28–30
  general tab, 32
  Navigation tab, 31
Creating a tour from a folder, steps, 109
Crime analysts, 2, 43
Criminal court, 3
Custom icons, 79, 89
  Map Icons Collection, 79
  useful sites, 79
    Benjamin Keen, 79
    The Google Developers site, 79
    Mapito, 79
Customizing Placemarks, 88
Custom maps, 4

## D

Defense Navigation Satellite System (DNSS), 35
Differential GPS (DGPS), 41
Digital forensics, 2, 3, 69
  process of, 69
    acquisition stage, 71–72
    analysis stage, 72
    reporting, 72–73
    seizure, 70
Digital image properties, 46
Distributing work, in Google Earth, 109
Document Type Definition (DTD), 58
2D triangulation, 38
Dual Frequency measurements, 40
3D viewer, 41, 42, 43, 97, 100, 105, 106
  changes, 20
  GIS dataset, 44
  photo, overlay, 98
  tab, 29
  trouble shoot, 30

## E

Email Program, 33
Enable placemark balloon local access, 80
Enabling access to local files, 80
ESRI file, 6
  ArcGIS, 44, 45
  ArcMap 10, 45
  shapefiles, 43
European Galileo, 36
Exchangeable image file format (Exif), 45

## F

File converters, 76
  GPSBabel, 76
  GPS Data Team, 76
  GPS Visualizer, 76
  TraceGPS, 76
Fingerprints, 2
Firefox, 5
Flavors, 4
  desktop, 4
  enterprise, 4
  mobile, 4
  Web, 4
Forensics, 2. *See also* Google Earth Forensics
  digital, 3. *See also* Digital forensics
  Google earth for, 2
  GPS device, 73
  HTML, 51
  tool, 3

Forensics *(cont.)*
create maps that display locations, 3
determine the location, 3
import data from mobile devices and GPS, 3
Free *vs.* Pro Desktop Versions, 6–8

## G

GE. *See* Google Earth (GE)
GEE. *See* Google Earth Enterprise (GEE)
Geocoding, 47
Geographic Information Systems (GIS), 6, 42–43
ArcGIS ArcScript, 45
usages, 42
Geo-location information, 3
in pictures, 45–47
Geotagging, 45, 47
information on digital image, 46
GeoTIFF files, 43–45
GIS Data
converting into KML, 45
and Google Earth, 43–45
Google Earth Pro to import, 43
Global Positioning System (GPS), 35
components, 36
control segment (CS), 37
space segment (SS), 36
User segment (US), 37
data, 73
data from a garmin imported into Google Earth, 87
differential GPS (DGPS), 41
error correction, 40
error solutions, 40–41
GPS Forensics, 73–74. *See also* Forensics
tools for recovering evidence, 74
Write Protection, 75
groupings, 36
importing GPS Data into Google Earth, 41–42, 88
measuring distance, 39–40
project, 35
trilateration, 37–39
Google, 4, 47
Google Earth (GE), 1
features, 1
usefulness, 1
investigating crimes to sharing information with, 2
resolve potential land-claim issues, 2
track environmental changes, 2
U.S. Fish and Wildlife Service provides data on wetlands, 2
Google Earth Enterprise (GEE), 5
components, 5
Client, 5
Fusion, 5
Google Earth API, 5
Server Software, 5
enterprise version, 6
usages, 6
advertising company, 6
police department, 6
Google Earth Forensics
practical application of, 83–85
scenario, 83
smartphone, lab for processing, 84
Google Earth Pro, 6, 9, 43, 45
importing Raster Data into, 44–45
importing Vector Data into, 44
installing, 9
link, 8
Google Earth toolbar, 14
Google Maps, 5, 47, 82
Google Play, app stores, 5
GPS. *See* Global Positioning System (GPS)
Guidance Software, 76

## H

HTML (Hypertext Markup Language), 50
code, 52
description tab of new placemark dialog, 51
document, 50
in google earth, 51–52
in Placemarks, 65–67

## I

Image Overlays, 96
Importing GPS files into Google Earth, 87
Indian Regional Navigational Satellite System, 36
Installing the Pro version, 8
Internet, 8
Ionosphere, 40
iOS, 5
IT department, 3
iTunes, app stores, 5

## J

JavaScript, application programming interface, 4

## K

Keyhole Earth Viewer, 53
KML (Keystone Markup Language), 50, 53
Basic KML Tags, 60
Error Handling, 33
file, 5, 28, 45, 53, 88, 95
example of, 55
opening files, 54
saving files, 53–54
viewing code in, 55–56
hacking, 103
placemarks, 60
revisited, 59
tags, 60
template file, 103
KMZ files, 28, 54

## L

Legend, applied, 103, 105
Location-specific information, 3
Location with trail layer activated, 92
Logos, applied, 103, 106

## M

MapInfo files, 10
tab, 6, 43
Markup languages, 49
kinds of. *See* HTML; KML; XML
load Web page in Internet browser, 50
logic, 50
Text between body and /body, 50
Text between em and /em, 50
Text between html and /html, 50
Text between p and /p, 50
Text between strong and /strong, 50
tutorials, 68
Master control station (MCS), 37
Measuring, 99
distance from satellites, 39
polygons and 3D buildings, 99
Metadata tags, 45
Misinterpreted Content, 81
Mobile phones, 3

## N


Navigation, 19
  looking, 20
    from a fixed vantage point, 20
  mouse controls, 8
  systems, 35
  through 3D maps, 4
  tools, 21
    Keyboard Shortcuts, 22
Navstar-GPS, 35
New Placemark dialog box, 17
  fields and tabs, 17–19
    altitude, 17
    description tab, 17
    latitude and longitude fields, 17
    name, 17
    style, color, 17
    view, 17


## O


Operating systems, 4
  Apple Mac OS X 10.6, 4
  Microsoft Windows Vista, 4
Organizing case, 77
  adding a folder, 78
  custom icons, 79
  enabling access to local files, 80


## P


Panoramio, 47
Pegman icon, 23
Placemark altitude
  adjusting, 91
Placemarks, 60–64
Plug-in, 3D globe, 4
Points of interest (POI), 14
Projection file (.pri), 44
Pseudo Random Code, 39


## R


Raster datasets, 43
Recording Voice Over Narration, 109
Removing Content, 82
Ruler dialog box, 99
Russian GLONASS, 36


## S


Saguaro National Forest folder, 107
Satellite images, 4
Server machines, 5
Shawnee Police Department, 2
Slider control for viewing historical imagery, 25
Smartphone, 5, 45, 92, 97
Style, color tab of placemark properties, 89
System requirements, 8
  high-speed Internet, 8
  operating system, 7
    Linux (Kernel 2.4or later, with 2.6), 7
    Mac OS X 10.6, 7
    Windows XP, 7


## T


Tablet, 3, 5
Tilting View, 20
Tools for recovering evidence, 74
  write protection, 75, 85
Tools used, to acquire evidence, 75
  cellebrite, 76
  file converters, 76
    GPSBabel, 76
    GPS Data Team, 76
    GPS Visualizer, 76
    TraceGPS, 76
  Guidance Software, 76
Touring settings, 107
Tours
  Guide, 26
  recording tours, 27–28
Traffic layer, 16
Traffic Near, 5
Trilateration, 37–39
Troposphere, 40


## U


US Daily Traffic Counts layers, 14
US Department of Defense, 36
User interface (UI), Google earth, 11, 12
  3D viewer, 12
  Layer panel, 12, 14
    display current or recent information about an area, 14
  Navigation controls, 12
  overview map, 12
  Parcel Search (APN), 13
  Places panel, 12, 16
    creating a new placemark, 18
  Scale legend, 12
  Search panel, 12, 13
  Sign in, 12
  Status bar, 12
  Toolbar buttons, 12–15
U.S. GPS, rivals to, 36


## V


Vector datasets, 43
Victim's parked Lincoln SUV placemark, 92
Viewing Historical Imagery, 24–25
Views, 22
  and camera angles, 101
  3D Views, 23
  in Google Maps, 25
  Ground View, 23
  Street View, 24
  tab of placemark, 102


## W


Waypoints, 35, 73, 74, 85
Weather layer, 16
Web developers, 4
Web pages, 4
Web requirements, 4
Web version, 4
Write blockers, 75
Write protection, 75
  software, 75
    DSI USB Write Blocker, 75
    M2CFG USB Write Block, 75
    NetWrix USB Blocker, 75
    Thumbscrew, 75
  tools for Mac computers, 75
    Disk Arbitrator, 75
    Softblock, 75


## X


XML (Extensible Markup Language), 50, 56
  basic features, 57–58
  code, 56–57
  rules for well-formed XML, 59
  syntax, 58


## Z


ZIP format, 54
Zooming, 19